# LAND USE
## Reflection on Spatial Informatics, Agriculture and Development

## About the Editors

**Mrityunjay Mohan Jha** received his Masters in *Geography,* M. Phil and Ph.D. in *food security* from the Jawaharlal Nehru University, New Delhi, besides receiving training in *Geographic Information System* from the Indian Institute of Remote Sensing, Dehradun. He has worked in the areas of decentralized planning, natural resource management, social assessment and resettlement and rehabilitation (R&R). His expertise also includes imparting training in participatory rural appraisal and social impact assessment. He has been part of research/consultative group for Department of Science and Technology, Ministry of Rural Development, Ministry of Human Resources Development, Government of India, IFPRI South Asia, ADRF and World Bank. He received an Excellence Award for 'people-centred work' from the Centre for Land Use Management, Hyderabad in 2005. Presently he is associated with the Intercontinental Consultants and Technocrats Pvt. Ltd., New Delhi. To his credit, he has three books and about a dozen research papers published.

**R.B. Singh** is presently Secretatary General, National Association of Geographers, India, and Reader in Geography, Delhi School of Economics, University of Delhi, India. He has specialized in Mountain Studies, Environmental Studies, Land Use/Cover Change, Disaster Management, Remote Sensing and GIS Technology. He has to his credit 25 research volumes/books and 145 research papers published in national and international journals of repute. The UNESCO/ ISSC (Paris) awarded him Research and Study Grants Award in Social and Human Sciences, 1988. Dr. Singh was trained in Remote Sensing and GIS Tech. in the field of Environment under GRID-UNITAR Programme. He has completed prestigious collaborative projects with ICSSR-IDPAD, CIDA-SICI and DFID. He is President of Environmental Monitoring Society. Earlier, he was President of Geography section of Indian Academy of Social Sciences and Recorder, Earth Science Section, ISCA. He is Steering committee Member of the IGU Commission on LUCC since 1996.

# LAND USE

## Reflection on Spatial Informatics, Agriculture and Development

*Edited by*

MRITYUNJAY M. JHA
R.B. SINGH

CONCEPT PUBLISHING COMPANY, NEW DELHI-110059

**Cataloging in Publication Data—DK**
*Courtesy:* D.K. Agencies (P) Ltd. <docinfo@dkagencies.com>

**Land use** : reflection on spatial informatics, agriculture and development / edited by Mrityunjay M. Jha, R.B. Singh.
p. cm.
Contributed articles.
With reference to India.
Includes bibliographical references.
Includes index.
ISBN 13: 9788180694639
ISBN 10: 8180694631

1. Information storage and retrieval systems—Land use—India. 2. Land use, Rural—India. 3. Land use, Urban—India. 4. Agriculture—Economic aspects—India. I. Jha, Mrityunjay Mohan. II. Singh, R. B. (Ram Babu), 1955-

DDC 333.731 309 54 22



ISBN: 13-978-81-8069-463-9
ISBN: 10-81-8069-463-1

First Published 2008



*Published and Printed by*
Ashok Kumar Mittal
**Concept Publishing Company,**
A/15-16, Commercial Block, Mohan Garden,
New Delhi-110059 (INDIA)

*Phones* : 25351460, 25351794
*Fax* : 091-11-25357103
*Email* : publishing@conceptpub.com

# Contents

# Acknowledgements

This book has taken shape as a result of interactions with professionals belonging to diverse fields, having interface with land use. We are grateful to Dr. K.V. Sundaram, who has been inspiring professionals towards achieving sustainable development and sustainable life styles.

We gratefully acknowledge Dr. K.D. Singh, former Chief Technical advisor, FAO, who has not only contributed to this book, but also constantly motivated us in completing this Volume. We express our gratitude towards Dr. N.C. Gautam, Director, Center for Land Use Management, Hyderabad, Dr. P.S. Roy, Deputy Director, National Remote Sensing Agency, Mr. M. Moni, Deputy Director General, National Informatics Centre, Prof. H. Ramachandran, Delhi School of Economics, Delhi University and Professor B. Thakur of Delhi School of Economics, for constant encouragement.

This book certainly would not have become a reality without the contributions made by eminent academicians, researchers and other professionals, who have contributed to this Volume. We acknowledge all of them.

We owe special thanks to the Concept Publishing Company, for bringing out this book, despite being flooded with other commitments.

# Introduction

Land Use as a subject matter has been widely discussed, debated and practiced, having interface with several disciplines and has remained in the center of planning in most of the developing and developed economies. Land as the most precious resource has only gathered strength, as numerous countries have moved from the subsistence mode to the intensive mode of land use. It connects us directly with land cover, biodiversity, agriculture, ecosystems, urban influence, economic development, land use planning, biogeochemical cycle, global warming and climate, among others.

The development of civilisation has been accompanied by radical changes in terrestrial environments. Although these changes have been successful to some extent, in providing food, fuel, and fibre for the ever-growing human population, they have been achieved at considerable environmental cost. One such cost is irreversible change in land use, which has led to soil and land degradation, major depletion in vegetation, decimation of plant and animal species, and worldwide reduction in biodiversity. Such situations have brought land use planning into focus, which varies according to contexts, meaning 'the scientific, aesthetic, and orderly disposition of land, resources, facilities and services with a view to securing the physical, economic and social efficiency, health and well-being of urban and rural communities'.

Conversion of land from one use into another may be reversible in some cases but not in others. Loss of elite soils to housing or urban development, is irreversible. Changes in land use impact directly on the ecosystem and are intimately linked with the issue of sustainability, and also have important consequences for food security. Understanding land use change in relation to its driving forces provides essential information for land use planning and sustainable management of resources.

The growing pressure of population coupled with increasing variety of demands made on land resources, have brought extra pressure on the available land resources. Hence, in order to use land optimally, it is not only necessary to have the information on existing land use/land cover but also the capability to monitor the dynamics of land use resulting out of changing demands of increasing population. World population increase, 2,40,000 people a day, pose immense problem to land resource management, augmenting problem of landlessness, land degradation and food security. Sustainability, the combination of production with conservation of resources to meet the needs of future generations, has often been recommended as key to land management.

Some major issues in policy and planning center upon land use, the availability of arable land, loss of forest cover and land needed to meet the urban expansion. This phenomenon is mostly associated with the developing world. Land use classification has been widely studied and researched by FAO and other national and international bodies, as a prerequisite for monitoring land use changes. The identifying and monitoring land use changes, using remote sensing and Geographical Information System is now effectively utilized for planning purposes. The classification of land for recommended uses depends fundamentally on an appreciation of human experience. Land resource, agriculture and development processes are closely linked in different ecosystems. Tropical ecosystem and biodiversity have raised concern among conservationists in the recent past due to increasing human interventions. Over utilization of forest resources has resulted in its depletion. Habitat loss and fragmentation of the natural landscapes have been found to be the main agent in the loss of biodiversity. Fragmentation generally results in the remnant areas of native vegetation surrounded by a matrix of agricultural or other developed land.

In this context, development of any country depends on the effective and optimum utilization of its various natural or physical resources. As early as in the 1960s, India realized the importance of remote sensing technology and its role in natural resources mapping and monitoring. Consequently the Indian Remote Sensing program was initiated in 1970. Modern remote sensing technology with its new detectors and imaging technologies provides a number of efficient and consistent ways to gather timely spatial data of the entire globe, at scales appropriate to a variety of applications.

Spatial Informatics (A combination of remote sensing and GIS), offers great opportunity in formulating a developmental plan at national, state and micro levels. This paves the way for the planner to acquire the required information at all levels with accuracy that can be used for various developmental activities and planning of existing resources in conjunction with the ground truth surveys. It also helps to survey various natural resources like land use, agriculture, forestry, minerals, water etc. and study various physical phenomena. The ability of GIS to identify new relationship from the existing database elements and to present the results of the analysis in cartographic form makes them exceedingly important from planning point of view, as most macro and micro-level planning tasks ideally require geographically referenced information. In India, great significance is attached to district level planning as the districts are to plan for their development considering their own Agro-climatic and socio-economic conditions and while doing so, the application of spatial informatics is capable of helping the planners in the areas, viz. conservation of resources, increasing production, reducing unemployment, alleviating poverty and environmental sustainability, among others.

This book is an outcome of constant interaction between academicians, scientists and researchers on certain topics revolving around land use. The ninteen papers included in this volume are contributed by professionals, at different stages of their working career, and belonging to different walks of life. Thematically, the papers fall under four broad categories, *viz. Application of Spatial Informatics in Land Use Studies, Agricultural Land Use and Development, Urban Influence and Land Use Dynamics and Socio-economic Development Linked with Land Use.*

**Application of Spatial Informatics in Land Use Studies**

At the global level, methods for the assessment of forest cover have been developed by FAO, explaining the salient features of the Tropical Forest Cover Survey, using the World Reference System-2 (WRS-2) of the LANDSAT satellites. The use of interdependent image interpretation is quite distinctive developed by FAO, provides forest cover change data of much higher precision than independent two date assessments. We also have an account of different methods of land use methods, highlighting the inevitability of remote sensing applications for depicting and measuring land use changes. Under a national scientific initiative India has developed comprehensive database on forest cover and biological richness at landscape level. In recent years, micro-level planning and decentralization has acquired greater significance, hence block and village level plans are of vital importance, and at these spatial levels, information provided by the satellite remote sensing has the capability to guide the local level planners, if effectively utilized.

## Agricultural Land Use and Development

Land use has remained dynamic around bigger settlements and urban centers, whereas in the large tracts of rural areas changes were not prominently visible. Such changes can be attributed to the influence of human being. The linkages between the human and land management process and drivers to land use need to be sufficiently understood. The complexity of dealing with the considerable variations in the drivers, and their consequences at the local, national, and regional levels, indicates towards the diverse issues that are associated with agricultural land use. The study of agricultural land use and productivity for countries facing limits to agricultural growth brings out food security issues. In such countries, sustainability remains the prime focus of their development process. Combined with tremendous demographic pressure, the questions of cropping intensity, physical and economic access to food and maintaining the productivity of land have gathered greater attention. Some of the other issues, which have remained in the periphery of debate and discussions, include impact of human resources on agricultural development, consequences of wars and local and regional conflicts on agricultural land use.

## Urban Land Use Dynamics and its Influence

Urban dynamics is the study of land transformations occurring in any urban areas through time. Urban areas in developing countries face the growing problems of haphazard growth, loss of vegetation, decline in arable land, and also decline in wildlife habitat. Most of the problems are attributed to increasing population pressure. The understanding of land use change in urban areas is key to the urban development programmes. The temporal analysis provides in any given context, the spatial patterns of land use. The emerged dynamism helps planners delineate urban growth boundaries, land use zoning plans and infrastructure requirements. Urban expansion and land reorganization has led to re-examining the socio-economic urban structure model, transformation of urban land use pattern, dipping levels of groundwater, increasing contamination of groundwater leading to environment and health hazards. In the developing countries, the impact of metropolisation on urbanised villages has thrown up challenging questions about land use dynamics, besides water facilities and sanitation. Urban expansion mostly engulfs its surrounding areas and acts as the facilitator of drastic changes in land use of its vicinity. Land use changes in the hilly areas in general and hill towns in particular have led to several insoluble problems related to environment and economy, exposing the development related problems and constraints.

## Socio-economic Development

Socio-economic development is inextricably linked with the current as well as past land use. In fact, land use pattern and access to land determine the socio-economic development of vast stretches of rural areas. Village development is often connected to land use and type and intensity of agricultural practices. For socio-economic development, an integrated approach is recommended for sectoral planning and management of land and its resources. The driving forces of historical as well as present land use changes, which transformed many societies and socio-economic conditions can be categorised into socio-economic, biophysical and land management themes. These forces include use of labour in the past history, national and international policies, i.e. access to credit, taxes, subsidies, conventions, and agreements, market forces, national and global demographic and social changes, environment, climate, pests and diseases, technological changes in transportation systems and mechanisation, parcellisation, subdivision, land values, agricultural diversification, urbanisation and industrialisation, changing cropping patterns, use of wetlands, waste generation, disposal and landfills, etc.

The distinctive purpose of this book is to provide different facets of land use and offer approaches to understand the multidimensional nature of land use. All the material in the book is arranged into chapters consistent with the dire need for holistic categories. This valuable compilation will be of immense use for all those who are interested and concerned with land use in developing countries. We express our gratitude to the Concept Publishing Company, New Delhi, for bringing out this Volume.

**Mrityunjay M. Jha**
**R.B. Singh**

# List of the Contributors

**Bomjan, Sushma** Lecturer (*Ad-hoc*), Shaheed Bhagat Singh College, University of Delhi, Delhi.

**Dabhadkar, Kaveri** Reader, Government Women's College, Bilaspur.

**Facknath, Sunita** Faculty of Agriculture, University of Mauritius, Mauritius.

**Fazal, Shahab** Lecturer, Department of Geography, Aligarh Muslim University, Aligarh.

**Gautam, N.C.** Director, Centre for Land Use Management, Hyderabad.

**Guleriya, Amar S.** Head, Department of Economics, Himachal Pradesh University, Summer Hill, Shimla.

**Jain, Jawahar Lal** Professor of Geography, H.S. Gour University, Sagar.

**Jha, Mrityunjay M.** Intercontinental Consultants & Technocrats, New Delhi.

**Joshi, P.K.** Scientist 'C', Forestry and Ecology Division, Indian Institute of Remote Sensing, Dehradun.

**Kareriya, B.R.** Lecturer, Department of Geography, Butwal Multiple Campus, Butwal, Tribhuwan University, Nepal.

**Lalljee, B.,** Faculty of Agriculture, University of Mauritius.

**Madhavi, Jupaka** Research Scholar, Jawaharlal Nehru University, New Delhi.

**Marh, Bhupinder S.** Chairman, Department of Geography, Himachal Pradesh University, Summer Hill, Shimla.

**Rathore, P.S.** Manager (Projects), ADCC Infocacl Pvt. Ltd., Atrey Layout, Nagpur, India.

**Roy, P.S.** Dy. Director, National Remote Sensing Agency, Department of Space, Government of India, Hyderabad.

**Sharma, Kaushal K.** Reader, Department of Geography, Kirori Mal College, Delhi University, Delhi.

**Singh, Anju** Lecturer, Aditi Mahavidyala (University of Delhi), Bawana, Delhi.

**Singh, K.D.** former Chief Technical Advisor, FAO, Rome.

**Singh, R.B.** Reader, Department of Geography, University of Delhi, Delhi.

**Singh, Rameshwar Prasad** Sr. Lecturer, Bhim Rao Ambedkar College, University of Delhi, Delhi.

**Singh, S.B.** Ex-Head, Department of Geography, Banaras Hindu University, Varanasi.

**Singh, Shyam B.** Research Scholar, P.G. Department of Geography, Utkal University, Vani Vihar, Bhubaneshwar, Orissa, India.

# PART I

# Application of Spatial Informatics

# 1

# Methodology and Results of FAO Survey of Tropical Forest Cover and Change Processes Based on Multi-date High Resolution Satellite Data

K.D. SINGH

## Introduction

For nearly 50 years (1946-2000), the Food and Agriculture Organization (FAO) of the United Nations has assessed the state of the world's forest resources at 5-year intervals during 1946-63 and at 10 years interval during 1970-2000. The main source of information has been the data available with countries. These assessments have been the only and very valuable source of forest resources information at the global level used by scientists, media persons and decision-makers.

The Stockholm Conference on Human Environment, 1972, brought the on-going tropical deforestation into sharp focus. Since then, many questions have been asked by the international community on the state of tropical forests such as:

- How fast are the tropical forests depleting?
- How much of forest being degraded? Fragmented?
- What is happening to deforested land?
- How are forest state and change linked with ecological and socio-economic conditions?
- What are the causes of deforestation?

Utilizing the advancements in satellite remote sensing and GIS, FAO launched (as a part of 1990 round), a major initiative to make a remote sensing based-continuing assessment of tropical forest cover and, more important to provide reliable information on the path of change. The survey belongs to the class of designs termed as "continuous forest inventory"; known to be efficient for providing change information. The specific objectives of the survey were to:

- Achieve the highest possible level of consistency and precision in the assessment of forest cover state and change at global and regional levels;
- Develop and disseminate a simple and robust monitoring technique for producing the forest cover state and change estimates at global and regional levels with application also at national level; and
- Provide spatial and statistical data for estimating class to class change of land cover and forest cover categories between the two dates of interpretation at the sample locations and for producing change matrices at regional and global levels.

Keeping in view the cost, precision and timeliness of results, a sampling approach based on remote sensing was used. Though applied only in the tropical zone, the survey methodology is general and can easily be applied worldwide for all the forest formations on the extent and the process of change (MI a consistent and continuing basis). The method is termed as "interdependent image interpretation" procedure and can be equally applied in conjunction with wall to wall mapping at a national and local levels.

**Methodology**

The World Reference System-2 (WRS-2) of the LANDSAT satellites was used to construct the sampling frame. LANDSAT scenes covering approximately 3.4 million hectares serve as sampling units. In view of cost-benefit considerations, the sampling units have been selected from all LANDSAT scenes with a minimum land area of 1 million hectares and a forest cover of 10 per cent or more, estimated on the basis of existing vegetation maps. This has restricted the area surveyed to some 62 per cent of the total tropical land area but containing some 87 per cent of the total tropical forest. The main characteristics of the survey were as follows:

- It covered the entire range of woody vegetation formations in the tropics, from the rainforests of the wet zone to the shrub and tree savannas of the dry zone.
- It provided estimates of mean forest cover area and area change together with an estimate of associated error.
- It was cost-effective through use of two-stage stratification based on existing FAO global statistical and GIS database.

**Land Cover Classification Used in FAO 1990 Assessment**

| *Main Classes* | | *Additional Classes* | |
|---|---|---|---|
| **Closed Forest** | (canopy cover >40%) | **High density** | (canopy cover >70%) |
| | | **Medium density** | (canopy cover 40-70%) |
| **Open Forest** | (canopy cover 10-40%) | | |
| **Long Fallow** | (forest affected by long fallow shifting cultivation) | | |
| **Fragmented Forest** | (mosaic of forest/non-forest) | | |
| | | **Dense fragmented** | (forest fraction 40-70%) |
| | | **Sparse fragmented** | (forest fraction 10-40%) |
| **Shrubs** | | **Dense shrubs** | (canopy cover >40%) |
| | | **Sparse shrubs** | (canopy cover 10-40%) |
| **Short Fallow** | (agricultural areas with short fallow period) | | |
| **Other Land Cover** | | | |
| **Water** | | | |
| **Plantation** | (man-made woody vegetation) | **Forest plantation** | |
| | | **Agricultural plantation** | |

This first survey round was conducted during 1992-95 using stratified randomly chosen 117 sampling units, an intensity of nearly 10 per cent. The distribution of the selected sampling units by region was 47 in Africa, 30 in Asia and 40 in Latin America. This sample size was expected to provide estimate of forest cover at global level with a standard error less than ±5 per cent. The same sites have been visited again during 2002-2003 as a part of FAO Global Forest Resources Assessment 2000.

At each sample location, satellite images of the best quality and appropriate season, separated by an approximate ten-year interval, were selected for study. The image close to 1990 in the first round provided assessment of the current state; whereas the area in common between the "1990" and "1980" images provided the assessment to be made of the changes over time. For the 2000 round also, a set of two satellite images of 2000 and 1990 were interpreted for the same sites.

A distinctive feature of the methodology (viz. interdependent image interpretation) lies in the fact that it provides not only forest cover change data of much higher precision than independent two date assessments, but also maps and change matrices for each sample location. This enables estimations of class to class changes of land cover and forest categories between the two dates of interpretation at sample and its aggregation at the regional and global levels: this is essential information for understanding the complex processes such as deforestation, fragmentation, degradation and afforestation.

The salient features of the survey from a remote sensing point of view were the following:

- Standard classification of various forest classes (closed, open, with shifting cultivation, fragmented) on a pan-tropical basis;
- Interdependent interpretation procedure: this interpretation approach secures the highest level of thematic and spatial consistency between historical and recent image classification. This procedure is the most important element of the methodology since it reduces the error associated to the estimate of changes and makes production of change matrices possible (see Annex 1);
- Image archive: all images used represent permanent reference as part of a continuous time series; in future these images will be used to estimate the speed of change (3 or more time series);
- Low sophistication in spite of its sound conceptual basis, the methods and procedures developed for this survey are simple and robust and require very low technological inputs; the methodology has been designed for implementation in average developing country conditions;

- Flexibility: although applied here in a global survey context, the monitoring methodology can be applied at national or sub-national level without substantial modifications.

The interpretation was implemented at selected regional and national forestry and/or remote sensing institutions, which had a good knowledge of the sample locations and are traditionally involved in forest resources assessment activities. With the two-fold objective of strengthening national capacities for forest monitoring and improving the quality of image interpretation, the Project organized three regional workshops and eight training sessions with national institutions, benefiting 27 countries and 81 participants.

The results and quality of the interpretation by local institutions were centrally reviewed and evaluated. A database was established and analyses carried out as is presented in the following sections.

The 1990 Assessment Project's Geographic Information System (GIS), with its multiple layers of geo-referenced information, was an integral element of the survey both at the stage of survey design planning and at the stage of analysis of results.

**Main Results of the Survey**

The survey produced estimates of forest cover and deforestation rates, with associated standard errors, at pan-tropical and regional levels, for three different definitions of forest cover. The idea of using three definitions was to compare the important effect of such definitions on the estimation of deforestation rates. A most interesting result of the survey were set of transition matrices, one for each Sampling Unit, that describe in detail the class-to-class changes for the *land cover classes between two dates.* The matrices associated with the sampling units can be aggregated by geographic regions (such as Africa, Asia and Latin America) or by ecological zones (such as Wet and Very Moist, Moist, Dry) and statistics computed in each case following the standard statistical procedure. Mean transition matrices have been produced and subsequent change analysis has been carried out for the three regions and by three ecological zones. First, the transition matrix for **the global tropics** produced by FAO 1990 assessment is as follows:

**Table 1:** Pan-tropical Area Transition Matrix (1980-1990*)

| *Land Cover Classed in 1980* | | *Land Cover Classes in 1990 (Million hectares)* | | | | | | | | | *Total 1980* | |
|---|---|---|---|---|---|---|---|---|---|---|---|---|
| | | *Closed Forest* | *Open Forest* | *Long Fallow* | *Frag-mented Forest* | *Shrubs* | *Short Fallow* | *Other Land Cover* | *Water* | *Planta-tions* | *Million ha.* | *%* |
| Closed Forest | | 1275.91 | 8.97 | 9.27 | 9.17 | 2.53 | 21.57 | 34.79 | 1.78 | 3.95 | 1367.96 | 44.6 |
| Open Forest | | 0.86 | 283.31 | 1.30 | 5.18 | 1.46 | 2.40 | 10.18 | 0.05 | 0.21 | 304.94 | 9.9 |
| Long Fallow | | 1.10 | 0.26 | 48.61 | 1.08 | 0.79 | 2.35 | 2.27 | 0.05 | 0.01 | 56.54 | 1.8 |
| Fragment Forest | | 0.58 | 0.63 | 0.63 | 159.33 | 0.45 | 1.41 | 11.40 | 0.25 | 0.39 | 175.06 | 5.7 |
| Shrubs | | 0.15 | 0.20 | 0.26 | 0.14 | 152.69 | 0.34 | 19.17 | 0.19 | 0.15 | 173.30 | 5.6 |
| Short Fallow | | 0.56 | 0.29 | 0.46 | 0.39 | 0.16 | 119.79 | 1.30 | 0.19 | 0.17 | 129.31 | 4.2 |
| Other Land Cover | | 0.71 | 0.70 | 0.26 | 1.35 | 1.94 | 2.03 | 834.23 | 1.58 | 0.44 | 843.26 | 2.75 |
| Water | | 0.14 | 0.02 | 0.01 | 0.05 | 0.01 | 0.07 | 1.46 | Stable water | 0.02 | 1.78 | 0.1 |
| Plantation | | 0.05 | 0.03 | 0.00 | 0.00 | 0.00 | 0.01 | 0.11 | 0.00 | 15.68 | 15.88 | 0.5 |
| **Total 1990** | **Million ha.** | **1280.06** | **294.41** | **60.81** | **176.69** | **160.03** | **149.97** | **920.91** | **4.09** | **21.03** | **3068.01** | |
| | **%** | **41.7** | **9.6** | **2.0** | **5.8** | **5.2** | **4.9** | **30.0** | **0.1** | **0.7** | | **100.0** |

*Results are based on the **visible common area** between recent and historical satellite images. Stable water (ocean, sea, lakes and rivers) has been exluded from the analysis of changes.

Closed forest was the land cover class undergoing the largest number of changes, comprising 51.6 per cent of all changes observed, at pan-tropical level, the period 1980-1990; this change in closed forest can be sub-divided by destination categories, as follows:

(a) 20.5 per cent changed into permanent agriculture, cattle ranching, water reservoirs, etc., with a complete loss of tree cover and woody biomass;
(b) 13.5 per cent into shrubs and short fallow agriculture which still represent deforestation but with some remaining woody biomass;
(c) 5.1 per cent into fragmented forest, which represents partial deforestation, with loss of approximately two-thirds of the original forest cover;
(d) 10.2 per cent into open forest or long fallow shifting cultivation which represent forms of degradation implying consistent loss of biomass (approximately half); and
(e) 2.2 per cent into agricultural or forestry plantations.

The remaining changes occurred in open forest (12.1% of total change), shrubs (11.5%) and fragmented forest (8.8%), the rest (16%) being distributed among the remaining five classes.

Over 90 per cent of all changes are negative, implying a loss of biomass, while only 10 per cent or less are positive, implying an increase of biomass.

Various changes observed have varying environmental implications, which, however, cannot be described through the simplistic *dichotomy forest deforestation.* In order to understand these complex land cover dynamics and develop effective preventive or corrective measures, detailed classifications and new analytical tools, such as change matrices and woody biomass flux diagrams, are essential.

The detailed analysis of changes occurring by three **tropical regions** shows that there is not only a quantitative difference among the three tropical regions in rates of change but that, there are also strong regional characters in the processes of change that indicate distinctive cause-effect mechanisms.

Forests in Africa show a total rate of change of 6.6 per cent, manifesting almost equally all phases of negative change *(deforestation to other land covered to other wooded land, fragmentation* and *degradation),* such changes appear to be mainly the effect of rural population pressure, characterized by the transition from *closed forest* to *short fallow* (subsistence farming) and progressive intermediate stages of depletion.

Forests in Latin America show a slightly lower total rate of change: 5.9 per cent, but caused almost exclusively by *deforestation to other land cover* (and therefore reaching a higher deforestation rate than Africa); this type of change appears to be the effect of centrally-planned operations such as government resettlement schemes, large-scale cattle ranching and hydroelectric schemes dominated by transition from *closed forest* to *other land cover.*

Forests in Asia show a much higher total rate of change: 11.3 per cent, which includes, as main categories of change, *deforestation to other land cover* and *to other wooded land, degradation and conversion to plantations* (agricultural and forestry). These changes are the result of two driving forces: rural population pressure and centrally-planned operations; the former comprises the intensification and expansion of shifting cultivation practices which show transition from *closed forest* to *long fallow* and to *short fallow,* the latter caused by government resettlement schemes and large plantation programmes which show transition from *closed forest* to *other land cover* and to *plantations.*

*The* analysis by **ecological zone** reveals important elements on (i) the intensity and distribution of forest area changes, and (ii) their environmental impact. The analysis of land cover changes by ecological zone shows clearly that, *in all geographic regions,* the forests of the moist zone have changed, i.e., have been deforested, fragmented, degraded, etc., with much higher intensity than the forests in the wet and dry zones. The percentage of pan-tropical forest area that has changed in the period 1980-1990 is estimated at 10.1 in the moist zone *versus* 4.8 in the wet zone and 4.6 in the dry zone. However, an indicative analysis of the biomass involved in such changes shows that, owing to the important differences in forest biomass density among the three zones and to the different change processes, the biomass loss in the wet zone is only slightly less than that lost in the moist zone and probably ten times more than that lost in the dry zone.

From the results summarized above, it would appear that socio-economic and cultural factors, which are more homogeneous within geographic regions, determine the nature of forest change processes while ecological characteristics determine the intensity of change.

FAO 2000 survey produced comparable results on a global tropical, regional and eco-region levels. Table 2 provides comparative data provided by 1990 and 2000 Assessments. As mentioned earlier, design is such that it can be replicated at suitable intervals to provide reliable information on State as well as trends.

**Table 2:** Forest Cover State and Change Assessments by FAO 1990 and 2000 Survey Rounds

*(All figures are in million of ha.)*

| *Region* | *Land Area* | *Forest Cover Estimates* | | | | | |
|---|---|---|---|---|---|---|---|
| | | *1990 Round* | | *2000 Round* | | *Change Estimates* | |
| | | *1980* | *1990* | *1990* | *2000* | *1980-90* | *1990-00* |
| Africa | 1224 | 584 | 556 | 542 | 519 | 28 | 23 |
| Latin America | 1234 | 878 | 824 | 805 | 780 | 54 | 25 |
| Asia | 611 | 316 | 290 | 316 | 272 | 26 | 44 |
| Tropics | 3068 | 1778 | 1671 | 1663 | 1571 | 107 | 92 |

The figures obtained for the common year 1990 in the two surveys is rather close, hardly differing by 8 million ha in a total of nearly 1600 million ha. However, regional level differences are more pronounced. It may be noted that at global tropical level, the survey shows a reduction in the rate of deforestation from 107 million ha during 1980-90 to 92 million ha during 1990-2000.

## Evaluation of Survey Precision

It may be noted here that "interdependent image interpretation" was a novel technique developed by FAO 1990 Assessment. Keeping this in view a statistical treatment of its superiority over commonly used independent assessment is given in Annex 1. The overall results of sample survey over the tropics is given in Table 3.

The statistical error associated with the global level estimates of forest cover at year 1990 for various definitions of forest cover, ranging between ± 3.5 and 4.5 per cent of the mean, is quite small and very close to the ± 3.9 per cent value estimated during the design phase of the survey. A higher margin of error is associated with mean deforestation rates, ranging between ± 12.3 and ±13.7 per cent of the mean rates. This is not too high, however, considering the high variance deriving from the 'event' character of deforestation

**Table 3:** Estimated Forest Cover, Deforestation Rates and Standard Errors at Regional and Global Levels for F-3 Definition.

| *Region* | *Forest Cover 1990 (%)* | | *Forest Cover 1980 (%)* | | *Deforestation rate (%) of 1980-90* | |
|---|---|---|---|---|---|---|
| | *Mean* | *Standard Error* | *Mean* | *Standard Error* | *Mean* | *Standard Error* |
| Africa | 45.44 | ± 3.22 | 47.72 | ± 3.32 | 4.77 | ± 0.61 |
| Latin America | 66.83 | ± 3.12 | 71.17 | ± 2.64 | 6.10 | ± 1.36 |
| Asia | 47.48 | ± 3.16 | 51.72 | ± 3.19 | 8.20 | ± 1.44 |
| Global | 54.45 | ± 1.90 | 57.94 | ± 1.81 | 6.03 | ± 0.75 |

Considering the extremely poor information available on the rate and processes of change of the tropical forest resources before the survey, the remote sensing component was an important achievement. A team of 10 independent peer reviewers appointed by the International Union of Forest Research Organizations (IUFRO) endorsed the soundness of survey design and utility of results. The concerned division of the Organization recommended FAO to undertake the survey on a continuing basis. Following this recommendation, FAO did continue remote sensing assessment as a part of 2000 assessment using identical design and methodology. In all likelihood, the survey will be continued in the future 2010 assessment and would provide the international community reliable information on the state and changes in the tropical forest cover on a continuing basis.

## The Enabling Conditions for the Survey

The tropical remote sensing survey was a true global cooperative effort planned and implemented by FAO Forestry Department, Rome, as a part of Forest Resources Assessment Programme. It was made possible by dedicated contribution of scientists from all parts of the world including among others Dr. R. Czaplewski and Professor B. Ranne in the field of statistical survey design and analytical models; and Professor D. R. Pelz for organizing IUFRO review of the design; and Mr. R. Baltaxe for the development of the monitoring methodology.

Many donors provided in kind and cash support including among others: Governments of Germany, the Netherlands, Sweden and the

United States, FAO unit for promoting Technical Cooperation among Developing Countries, the European Community, the U.S. Forest Service, the Swedish University of Agricultural Sciences, the University of Freiburg, Germany, and the Overseas Agronomic Institute, Florence, Italy, the Environmental Protection Agency of USA (EPA) and the National Research Council of Thailand.

The survey was designed and implemented by a team led by Dr. K.D. Singh, FAO, Rome, who conceived the survey design on the lines of continuous forest inventory technique and secured financial resources and collaboration for its design and implementation. Mr. R. Drigo made the major contribution to remote sensing component of the survey including development and dissemination of the interdependent image interpretation technique, review of interpretation for all sampling units, study of change processes and writing the present report. Mr. M. Lorenzini developed the Project's Geographic Information System and, in cooperation with Mr. A. Marzoli and Mr. G. Mu'Ammar contributed to the definition and stratification of the sampling universe through the combined analysis of geographic and database information. Mr. A. Hildeman and Mr. H. Simons significant contribution to the dissemination of the monitoring methodology through the organization of regional workshops and training sessions. Ms. E. Milani contributed to the production of the various maps that are shown in the present report on the basis of the Project's Geographic Information System. In addition, the survey benefited from the collaboration with an extended network on a world-wide basis of institutions and individuals, remote sensing centers, and participants to the regional workshops, who all contributed in the interpretation of remote sensing images, by providing source information, consultancies and expert help. The Forest Resources Assessment 1990 was a major global cooperative effort planned, coordinated and implemented by FAO Forestry Department. This global cooperative effort and contributions made this challenging idea possible and successful.

## REFERENCES

Czaplewski Raymond, (1996): Evaluation of FAO Remote Sensing Survey 1990 (unpublished report), FAO Rome.

FAO (1982): Tropical Forest Resources Assessment, FAO Rome.

FAO, Rome FAO (1993): Forest Resources Assessment 1990: Tropical Forest Resources, *Forestry Paper* 112 Global Synthesis, FAO, Rome.

FAO (1995): Forest Resources Assessment 1990: Global Synthesis, Forestry Paper 124 FAO, Rome.

FAO (1996): Forest Resources Assessment 1990: Survey of Tropical Forest Cover and Study of Change processes. Forestry Paper 130.

FAO (1998): Strengthening of Country Capacity for Planning, Assessment and periodic evaluation of Forests: Findings and Recommendations, FAO/SIDA Project, Rome.

FAO (2001): Global Forest Resources Assessment 2000: Main Report, FAO Rome.

Kotka III Report 2000: Expert Consultation on Global Forest Resources Assessment, The Finnish Forest Research Institute, Helsinki.

UN (1996): Indirect techniques for demographic estimation: Manual X, UN Population Division, (in form of a draft manuscript), UN, New York.

UNEP (1972): United Nations Conference on Human Environment, Environmental aspects of Natural Resources Management.

*ANNEXURE 1*

**Interdependent Interpretation Procedure**

**Step 1**

Interpretation of 1990 image on a transparent film overlay (1990 overlay).

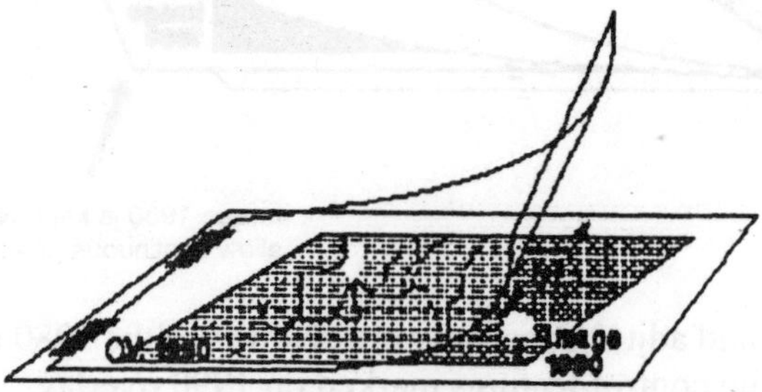

The interpretation includes the delineation of Land Cover Classes (black lines) and control features (green lines).

**Step 2**

Insert 1980 image between 1990 image and 1990 overlay. Use the green control features to adjust the position of 1980 image to fit the 1990 overlay.

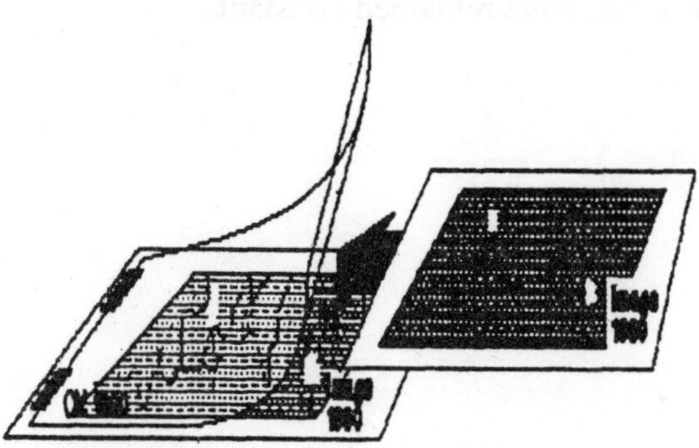

**Step 3**

Interpret 1980 image (in blue on 1980 overlay) using the delineation on the 1990 overlay as a reference.

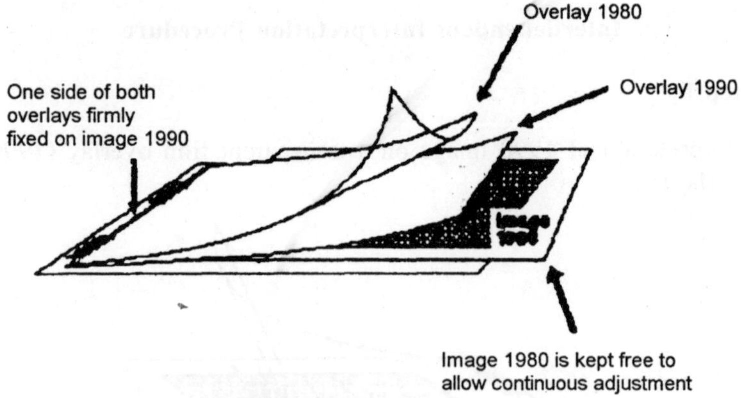


Check and adjust constantly the position of the 1980 image with respect to the control features marked on 1990 overlay.

The result of the interpretation process will be:

- Spatial co-registration of the two overlays according to the projection of the 1990 image.
- Complete interpretation of the two images without relative subjectivity in class delineation.
- Each line on each overlay is supported by the analysis of **both** images for the classes that have changed as well as for those that have remained constant.

*ANNEXURE 2*

## Comparison of Precision Associated with Alternative Image Interpretation Methods

Notation:

$X_1$ = Interpretation at date 1
$X_2$ = Interpretation at date 2
$Y_1$ = true class at date 1
$Y_2$ = true class at date 2

$\varepsilon_1$ = noise of image (date 1)
$\varepsilon_2$ = noise of image (date 2)
$\mu_1$ = error of interpretation (date 1)
$\mu_2$ = error of interpretation (date 2)

$(\varepsilon_1 + \mu_1)$ and $(\varepsilon_2 + \mu_2)$ are the error terms. As $\varepsilon$ and $\mu$ may be positively correlated; i.e., much noise in the image makes the interpretation more uncertain. Assuming independence between images, the total error is: $\text{Var}(\varepsilon_1 + \mu_1) = \sigma\varepsilon^2 + \sigma_\mu^2 + 2\text{Kov}(\varepsilon_1,\mu_1) = \sigma_e^2$

**Independent Interpretation**

$X_1 = Y_1 + \varepsilon_1 + \mu_1$ — Interpretation at date 1
$X_2 = Y_2 + \varepsilon_2 + \mu_2$ — Interpretation at date 2
$X_1 - X_2 = Y_1 - Y_2 + \varepsilon_1 - \varepsilon_2 + \mu_1 - \mu_2$ — Change estimation
$\text{VAR (state estimation)} = \sigma\varepsilon^2 + \sigma_\mu^2 + 2\text{Kov}(\varepsilon,\mu) = \sigma_e^2$
$\text{VAR (change estimation)} = 2\sigma\varepsilon^2 + 2\sigma\propto^2 + 4\text{Kov}(\varepsilon, \mu) = 2\sigma_e^2$

**Interdependent interpretation**

$X_1 = Y_1 + \varepsilon_1 + (\mu_1 + \mu_2)/2$ — Interpretation at date 1
$X_2 = Y_2 + \varepsilon_2 + (\mu_1 + \mu_2)/2$ — Interpretation at date 2

VAR **(state estimation where there is no change)** i.e., $Y_1 = Y_2$ Classification according to: $X = (X_1 + X_2)/2 = Y_i + (\mu_1 + \mu_2)/2 + (\varepsilon_1 + \varepsilon_2)/2$

$$\text{VAR}(X) = \sigma\varepsilon^2/2 + \sigma_\mu^2/2 + (2\text{Kov}(\varepsilon,\mu))/2 = \sigma_\varepsilon^2/2$$

VAR **(state estimation where the class has changed)**
Classification according to: $X_2 = Y_2 + \varepsilon_2 + (\mu_1 + \mu_2)/2$

$\text{VAR}\ (X_2) = \sigma\varepsilon^2 + \sigma_\mu{}^2/2 + \text{Kov}\ (\varepsilon_2, \mu 2) = \sigma_\varepsilon{}^2/2 + \sigma\varepsilon^2/2$
VAR (change estimation) = $\text{Var}\ (\varepsilon_2 - \varepsilon_1) = 2\sigma\varepsilon^2$

From the above formulation it would appear that the interdependent approach, where both images are used throughout the interpretation process, is beneficial not only in terms of change assessment but also in state assessment as the consultation of both images provides additional evidence in support of the delineation of the classes in both historical and recent images.

# 2

# Remote Sensing Applications for Land Use/Land Cover Mapping, Planning and Monitoring

N.C. GAUTAM

## Introduction

The growing pressure of population coupled with increasing variety of demands made on land resources, have brought extra pressure on the available land resources all over the country. Hence, in order to use land optimally, it is not only necessary to have the information on existing land use/land cover but also the capability to monitor the dynamics of land use resulting out of changing demands of increasing population.

Conventional ground methods of land use mapping are labour intensive, time consuming and are relatively infrequent. These maps soon become outdated with passage of time, particularly in a rapidly changing environment. In recent years, satellite remote sensing techniques have been developed, which have proved to be of immense value for preparing accurate land use/land cover maps and monitoring changes at regular intervals of time.

In case of inaccessible region, this technique is perhaps the only method of obtaining the required data on a cost and time-effective basis. An article on IRS-1A applications for land use/land cover mapping in India has appeared (*Current Science*, 1991 61, 153-161).

**Current Status**

The organizational efforts in publishing maps, reports and statistical data by Survey of India (SoI), National Atlas and Thematic Mapping Organization (NATMO), National Bureau of Soil Survey and Land Use Planning (NBSS & LUP), All India Soil and Land Use Survey (AIS & LUS), National Remote Sensing Agency (NRSA), Space Applications Centre (SAC), Central Arid Zone Research Institute (CAZRI), Ministry of Agriculture, Settlement Survey and Land Records, Revenue Departments, National Sample Surveys, State Land Use Boards, Town and Country Planning Organization (TCPO) and other local agencies are noteworthy.

Ever since the launch of the first remote sensing satellite (Landsat-1) in 1972, land use/land cover studies were carried out on different scales for different users. Wasteland mapping of India was carried out on 1:1 million scale by NRSA using 1980-82 Landsat Multi Spectral Scanner data. About 53.3 Mha (16.2%) of wastelands were estimated based on this study. Subsequently, mapping on 1:50,000 scale was carried out by NRSA for the entire country using remote sensing techniques. About 6,38,515 sq. km. (20%) of waste lands were estimated. These maps give five layers of information, e.g. base information, village, notified forest—micro watershed boundary and wasteland information upto village level.

At the instance of Planning Commission, the Department of Space carried out land use/land cover mapping of all 442 districts using IRS LISS-I data based on visual and digital interpretation techniques in collaboration with various work centers in different states. This task provided first national level spatial maps on land use/land cover along with statistics in 22 different categories. Since the mapping was carried out on 1:250,000 scale using a coarse resolution satellite data, small settlements, linear features like transportation network could not be delineated. To derive accurate statistics on agricultural land use, thumb rules were worked out based on extent of small settlements, length of transportation lines like roads and railways from the existing topographical maps for applying necessary corrections. Due to the small scale and two resolutions of the data used, the maps/statistics generated are moderately accurate and the statistics on land use categories could not be generated at village level.

The launch of IRS-1C and IRS-1D satellites with WiFS, LISS-III and PAN data have provided an opportunity for the first time to derive the land use/land cover information at regional, local and cadastral levels, on appropriate scales with an assured frequency for updation.

**IRS-1C/IRS-1D:** IRS-1C was launched on 28th December, 1995 and IRS-1D on 29th September, 1997, both the satellites have some sensors and other characteristics.

The WiFS data have the potential to monitor the dynamic phenomena every 5 days at level-I; LISS-III has the potential to generate the level-III upto 1:25,000 scale. The composite data of LISS-III and PAN have the potential to be enlarged on a 1:10,000 scale without loss of data to look out for cadastral level information. The possible scales on which land use/land cover can be generated, level of classification and the minimum mapable units that can be achieved using different sensors are shown in Table 1.

**Table 1:** Possible Scale and Minimum Mapable Units using IRS-1C for Land Use/Land Cover Analysis

| *Sensor* | *Scale* | *Minimum Mapable Unit (MMU)* | *Area (ha.)* | *Level Classification* |
|---|---|---|---|---|
| WiFS | 1:250,000 | 3 × 3 | 56.25 | Level-I |
| LISS-III | 1:25,000 | 3 × 3 | 0.56 | Level-II |
| PAN | 1:10,000 | 3 × 3 | 0.009 | Level-III/IV |

Realizing the excellent and useful information that can be extracted from IRS-1C & IRS-1D data, the Department of Agriculture, Government of Andhra Pradesh has requested NRSA to carry out mapping of five taluks to derive the statistics at village level which will most likely be extended to the entire state using LISS-III data.

## Objectives

The following are the objectives of this study:

- To analyse the land use/land cover using WiFS data for the State of Andhra Pradesh.

- To study the capabilities of LISS-III data and PAN and LISS-III merged data for large scale mapping at micro level of Kaziranga National Park in Assam.

**Study Area**

Area covered by Andhra Pradesh State was selected to analyze the WiFS data. Kaziranga National Park and environs in Assam were selected to analyze and derive the land use/land cover data using LISS-III. A part of Ranga Reddy district, Andhra Pradesh was selected for analysis of composite (merged) data of LISS-III and PAN data.

**Methodology**

***Interpretation of Remotely Sensed Data for Land use/Land cover***

A remote sensor records response which is based on many characteristics of the land surface, including natural and artificial cover. An interpreter uses the elements of tone, texture, pattern, shape, size, shadow, site and association to derive information about land use activities which is also the basic information about land cover. The title 'land use mapping' is quite often applied to remote sensing image classification procedure as a whole which tends to amalgamate the distinct concepts of mapping land use and land cover.

The generation of remotely sensed data/images by various types of sensors flown aboard different platforms at varying heights about the terrain and at different times of the day and the year, does not lead to a simple classification system. In fact, many researchers believe that no single classification could be used with all types of imagery and all scales. To date, the most successful attempt in developing a general purpose classification scheme compatible with remote sensing data show the Methodology (Fig. 1) for land use/land cover analysis attempted by Anderson *et al.* which is also referred to at USGS classification scheme. Other classification schemes available for use with remotely sensed data are basically modifications of the above classification scheme.

Four levels of classification are suggested in this scheme, each level being useful at various information/decision-making levels.

**Step by Step Methodology**

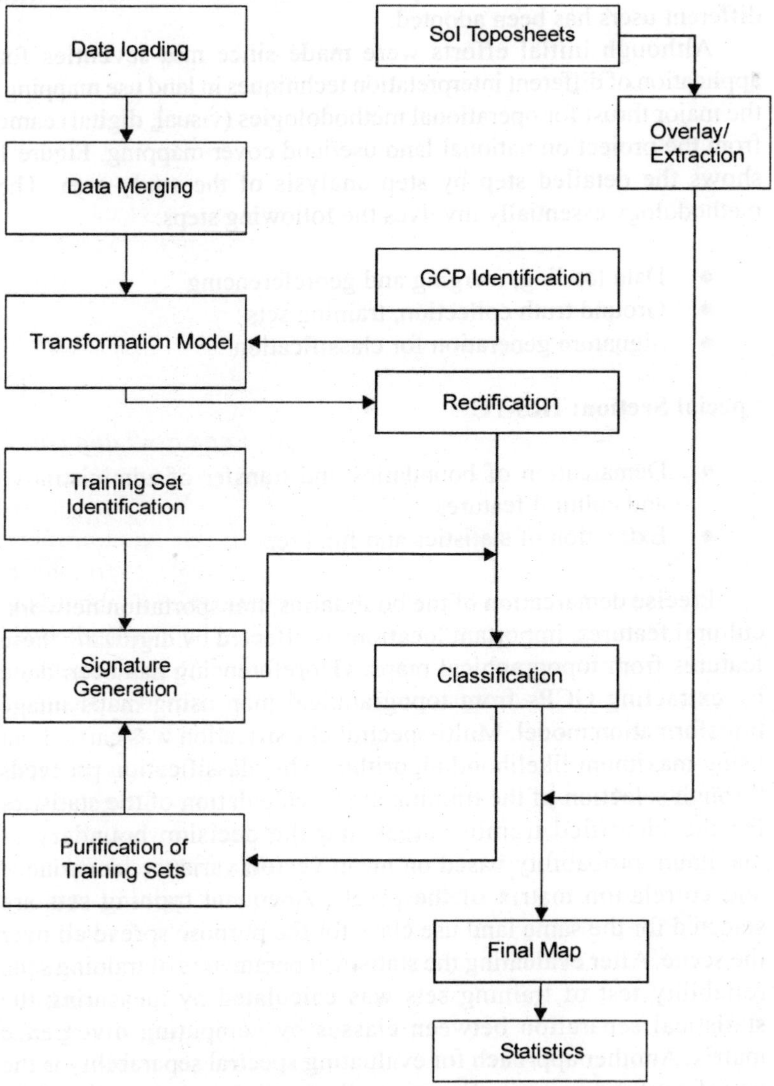


**Fig. 1:** Showing the Methodology for Land use/Land cover

Level-I information can be used at inter-state and state-level planning; level-II, at state-level to regional level planning, level-IV information can be used at local or micro level planning. In the present study, the classification system developed by NRSA in consultation with different users has been adopted.

Although initial efforts were made since mid-seventies for application of different interpretation techniques in land use mapping, the major thrust for operational methodologies (visual, digital) came from the project on national land use/land cover mapping. Figure 1 shows the detailed step by step analysis of the study area. The methodology essentially involves the following steps:

- Data loading, merging and georeferencing
- Ground truth collection, training sets
- Signature generation for classification

**Special Section: IRS-1C**

- Demarcation of boundaries and transfer of administrative and cultural features.
- Extraction of statistics and final report.

Precise demarcation of the boundaries, transportation network, cultural features, important locations is effected by digitizing these features from topographical maps. Georeferencing has been done by extracting GCPs from topographical map using maps image transformation model. Multi-spectral classification was carried out using maximum likelihood algorithm. The classification proceeds through selection of the training areas, calculation of the statistics for the identified training areas and the decision boundary of maximum probability based on mean vector, variance, covariance and correlation matrix of the pixels. Adequate training sets are selected for the same land use class for the purpose spread all over the scene. After evaluating the statistical parameters of training sets, reliability test of training sets was calculated by measuring the statistical separation between classes by computing divergence matrix. Another approach for evaluating spectral separability is the use of confusion matrix. Ensuring all non-diagonal elements of the confusion table as zero, misclassification can be minimized. The overall accuracy of the classification is finally assessed with reference to ground truth data.

## Results

### *Land use/Land cover Analysis using WiFS data*

About ten classes of level-I and level-II could be achieved using the digitally mosaiced WiFS data of Andhra Pradesh. Andhra Pradesh is covered in parts of three WiFS scenes (path/row 103-61; 103-66; 98-61). The digital analysis of land use/land cover has been carried out using three bands, viz., red, near infrared (NIR), and texture of NIR. The texture of NIR has been derived using textural analysis in EASI PASE software. The classified data of Andhra Pradesh are shown in Figure 2.

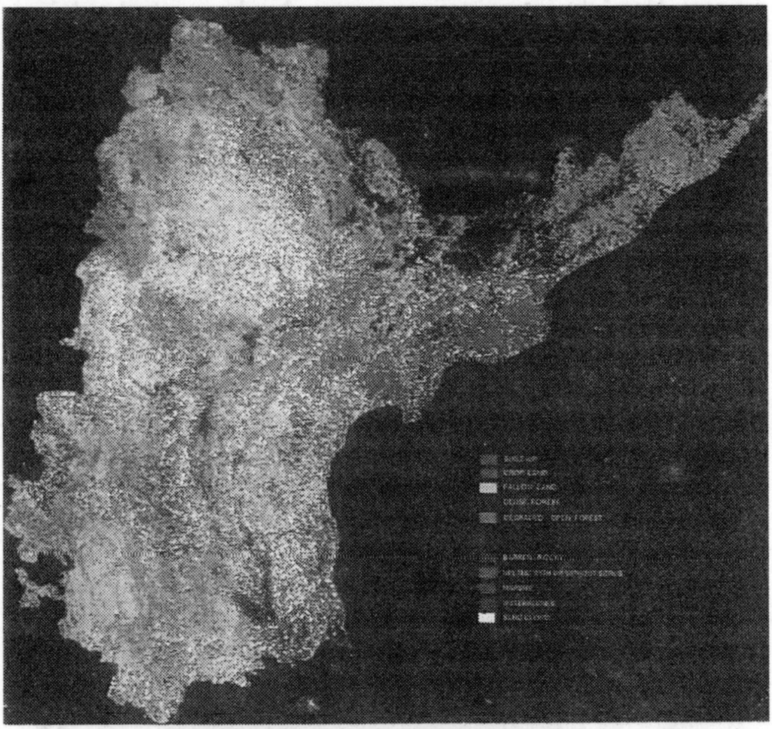

**Fig. 2:** Showing Classified Satellite Data of Andhra Pradesh.

WiFS data have been found to be adequate to bring out all the level-I classes derived from IRS-1A level LISS-I data on 1:250,000 scale under land use/land cover mapping of agroclimatic zone planning. In the present study winter season (rabi) data were used for analysis.

Due to the limitation of data (single season) some of the classes like open/degraded forest and uplands could not be separated in some areas, with the result that some of the uplands in parts of Nizamabad, Medak and Ranga Reddy districts were mixed with degraded forest. Also some of the fallow lands in south-western part of the state have been misclassified as upland due to similar spectral signature in both the cases. Using multi-season data, this problem can be solved by analyzing the accurate land use/land cover information. The land use classes like crop land, water bodies, barren lands, dense forest waterlogged areas could be identified and classified more accurately using WiFS data. The interpretation accuracy has been found to be as high as 80-85 per cent. Here, the interpretation accuracy has been calculated using the kappa accuracy, which is determined from the error matrix, considering the number of correctly classified units and also errors of commission and errors of omission.

***Land use/Land cover Analysis using LISS-III Data***

Digitally classified output of land use/land cover of Kaziranga National Park and the corresponding raw data are shown in Figures 3 and 4 respectively. The classification has been made using NDVI for areas inside the park and using maximum likelihood classification for the rest of the area.

**Fig. 3:** Showing Satellite Data of Kaziranga National Park.

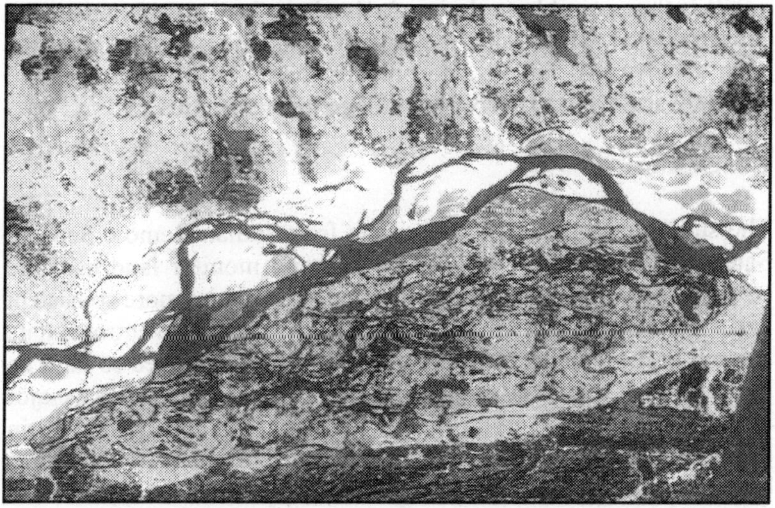

**Fig. 4:** Showing Digitally Classified Satellite Data of Kaziranga National Park.

The level-III classes such as type of plantation, viz., teak, sal, homestead plantations, different levels of grass in the Kaziranga National Park (High, Medium, Low) could be identified and classified. At the time of data collection, there were no crops in the study area and most of the agricultural area is under fallow. Major linear features such as roads, rail network and small settlements are clearly separable using LISS-III data. The water-logged areas within the national park are clear and areas of water bodies as small as acres also could be classified accurately.

***Land use/Land cover Analysis using Composite Data (LISS-III and PAN)***

The composite (merged) data of LISS-III using PAN of part of Ranga Reddy district, South of Hyderabad suggests that the land use/land cover details are very sharp in this data compared to LISS-III. Also all the linear features are clear and details like car tracks could be identified. The grape gardens along Nagarjunasagar road (regular, bright red patches), which are less than ½ acre extent are clearly seen. Individual farm houses and cluster of few houses are easily separable which is very crucial to derive accurate statistics of different land use categories.

It is thus possible to derive the land use/land cover statistics at cadastral level, which takes the process of mapping beyond what has been achieved so far, viz., to prepare spatial land use maps upto 1:10,000 scale and to derive the village level statistics.

## Conclusion

The study demonstrates that the data from Indian Remote Sensing Satellite (IRS-1C) can be used to map and monitor land use/land cover details rapidly at different levels. This provides more consistent and accurate base line information than any of the conventional data sources or the sources from any of the satellites launched so far. The availability of spatial information at village/cadastral level is a major step for planning at micro-level. Further, this data will help in estimating the area under different crops more accurately which forms base line information for crop production estimation on an operational basis.

## REFERENCES

Anderson, J.R. (1971): *Photogram, Engg.*, 37, 279.

Anderson, J.R. *et al.* (1976): *A land use/land cover classification system for use with remote sensing data*, USGS Publication, Professional paper 964.

Anonymous (1989): *Manual of nationwide land use/land cover mapping using satellite imagery Part-I*, NRSA, pp.1-58.

Anonymous (1990): *Manual of nationwide land use/land cover mapping using digital techniques, Part-II*, RRSSC, Nagpur, February, pp. 1-78.

Fitzpatrick, Lins, K. and Chambers, M.J., J. Rem. (1977): *Sen Electromagnetic Spectrum*, 4.41.

# 3

# Geospatial Tools for Forest and Biodiversity Conservation
## *Indian Perspective*

P. K. JOSHI AND P. S. ROY

## Introduction

The planet earth is unique in several respects—supporting lives, structurally complex, rich in resources, diversified into several subtypes and highly productive in terms of energy. There is now increasing awareness of land concern with the importance of natural resources. Initially natural ecosystems and their functioning were guiding factor for human activities in the biosphere. However, overriding priorities and the human activities modify the landscapes to suit their own needs. The human land uses dramatically alter the ecological character of the land where they occur, and they tend to be permanently modified at ecological time scales. The proximate and underlying drivers of human transformation of Earth face affecting ecosystem organization and functions are given in Table 1. Among the natural resources Tropical forests occupy less than 7 per cent of the terrestrial surface yet contain more than half of all plant and animal species (Mayers, 1992). Tropical deforestation is responsible for massive species extinction and affects biological diversity in three ways, viz., habitat destruction, isolation of formerly contiguous forest into forest fragments and adverse physical and biological consequences of edge and 'buffer' effects within a boundary zone between forest and deforested areas (Pimm *et al.*, 1985 and Prance, 1982).

**Table 1:** Proximate and Underlaying Drives of Human Transformation of Earth's Face Affecting Biological Organisation and Ecosystem Functioning.

| *Nature of Transformation* | *Proximate Drive* | *Underlaying Drive* |
|---|---|---|
| Land | Clearing (Cutting + Burning), agriculture practices; abandoning the land and subsequent degradation. | Demand for food, environmental quality, ecosystem goods and services, fragmentation leading to isolation of habitats/ecosystems. |
| Coastal Zone | Land cover conversion, groundwater exploitation, urbanisation. | Environmental quality, disappearance of fragile ecosystems, decline in livelihood condition. |
| Wetland | Landuse alteration, drainage change, desiccation, eutrophication. | Change in hydrological regimes and land surface characterisation, groundwater depletion, alteration in food chain and loss of keystone species. |
| Biodiversity | Clearing of forest/natural ecosystem, introduction of alien species, making landraces unpopular. | Risk to environmental and food security, landscape amenity and disappearance of habitats/ecosystems. |

The impact of any land use activity depends on (1) the degree of habitat modification and (2) the spatial extent and pattern of habitat modification. Spatial organization of the landscape connectivity is important for sustained ecological functioning. Connectivity can exert strong influences on ecological processes such as the movement and dispersal of organism (Gardner *et al.*, 1989), the use of resources by animals, i.e. corridors (O'Neill *et al.*, 1988a, 1988b), gene flow (Gilpin and Soule, 1986), and spread of disturbance. Changing the landscape patterns through fragmentation can disrupt ecological processes that depend on movement within the landscape (Roy *et al.*, 2002). Even climate change is likely to have considerable impacts on most or all ecosystems. The distribution pattern of many species and communities are determined to a large part by climatic parameters. However, the responses to changes in these parameters are rarely simple. Some of the impacts of habitat alterations in the critical ecosystems of tropics are given in Table 2.

**Table 2:** Critical Ecosystem/Landscape Types (http://www.unep-wcmc.org).

| *Ecosystem/ Landscape Type* | *Key climatic variables* | *Implications for biodiversity* |
|---|---|---|
| Wetlands | Mean summer temperature. Mean annual precipitation Flooding. | Increased variability in the hydrological cycle leaving inland wetlands to dry out with lower species diversity. Warming of 3-4°C could eliminate 85 per cent of all remaining wetlands. |
| Coastal marshes | Relative rate of sea-level rise, changes in hydrological balance Storm frequency and severity. | Habitat loss of estuaries and deltas, particularly where these are backed by agricultural or urban land, preventing natural retreat Implications on migratory species and their flyway patterns. |
| Forests (general) | Changes in rainfall, temperature and potential evapotranspiration. Increased frequency of fire and storms. | Major changes in vegetation types, forests may disappear in certain areas at a rate faster than the potential rate of migration to, or re-growth in, new areas. |
| Tropical Montane Forest | Changes in degree of cloud cover and sunlight hours. Hurricane frequency/ severity. Drought frequency/ annual rainfall. | Drying out and invasion or replacement of montane species by lower montane or non-montane species. |
| Arid and semi-arid areas | Precipitation patterns. Minimum winter temperatures. | Deserts are becoming more hotter and drier. Desertification into sub-Saharan Africa and Central Asian Steppes Salinisation. Loss of grassland/arable land. |
| Coral Reefs | Sea-surface temperature, indirect chemical effects of higher $CO_2$ concentrations in the water. | Prolonged exposure to minor (>1°C) rises in temperature causes coral bleaching/death. Impacts may be compounded by chemical effects of increasing $CO_2$ concentrations which could reduce calcification rates (skeletal development, growth rate). |
| Mangroves | Relative rate of sea-level rise, changes in hydrological balance Storm frequency/severity. | Decrease in extent as coastal zone becomes "squeezed" between sea and inland agriculture. |

**Biodiversity**

Biological diversity (commonly referred as biodiversity) is part of our daily lives and livelihood, and constitutes resources upon which the humanity depends. It is not uniformly distributed on the earth. Increasing human interventions on the ecosystems have accelerated the process of biodiversity loss. Though process of speciation and extinction is part of evolutionary process but it is reported that the speciation is more concentrated along tropics where the ecological set up and continental readjustments have permitted higher rate of speciation. The recent concern for loss of biodiversity and its analysis reveal that the species extinction process has overtaken speciation process (UNEP, 1995). An environment, which is rich in biological diversity, offers the broadest array of options for sustaining human welfare and for adapting to change. Loss of biodiversity has serious economic and social costs for any country. Conservation and sustainable use of biodiversity, hence, is fundamental to the sustainable development. The experiences of the past few decades have shown that as industrialisation and economic development patterns of consumption, production and needs change, strain, alert and even destroy ecosystems.

Biodiversity is the totality of genes, species, ecosystems and habitats in a region. The smallest unit starts from the diversity contained in the genetic material of individual organisms and then goes on to encompass the biological communities in which species are organized and on to the ecosystems in which they exist. Biodiversity is recognized at three levels, viz., genetic, species and ecosystem levels. At all three levels of biodiversity, the components are dynamic in space and time. A recent analysis of future biodiversity trends in the major biomes of the world identified the main cause of biodiversity loss in the coming decades as land-use change, mainly loss of habitat and landscape fragmentation.

Changes in atmospheric composition and climate are regarded as long-term factors, increasing in relative importance over time. However, changes in nitrogen deposition have important impacts in short timeframes on species diversity, especially in the developed world. Research on the consequences of these changes to biological diversity for the functioning of terrestrial ecosystems is in its infancy. Biodiversity is generally the greatest in the oldest ecosystems. It changes across environmental gradients, like latitude, altitude, depth, aridity etc. Any changes in the climatic variable will impact on biological system and their organization. The genetic composition of species changes over time in response to natural and human induced selection pressures, occurrence and relative abundance of species in ecological communities as a result of ecological and physical factors.

### *Biodiversity Loss Due to Forest Fragmentation*

Clearing of natural systems has created a highly modified landscape where remnant patches of native flora are set in a matrix of agricultural lands and urban-residential development (Holl, 1999; Turner and Corlett, 1996; Williams-Linera *et al.,* 1998). Prominently deforestation of tropical forests has been rapid and extensive; between 1981 and 1990, the total loss of natural tropical forests was 154 ha, cleared at a rate of nearly 1 per cent per year (Whitmore, 1997). Loss of biodiversity in the tropics is directly related to forest clearing, which decreases total habitat area (Dale *et al*., 1994). However, the effect of deforestation, changing landscape pattern, also negatively impacts of biodiversity. Fragmentation causes vegetation patches to be situated in different positions in the landscape, varying in size, shape, isolation (Saunders *et al*., 1990), and time since excision from the continuous forest (Turner and Corlett, 1996). These factors, in turn, influence the biodiversity of a fragment, "and in a complex manner, the biodiversity of the collection of fragments that occupies the landscape" (Turner and Corlett, 1996). Despite negative effects caused by fragmentation, conservation in the modified tropical landscape must assess individual forest fragment's contributions to the protection of biodiversity. To protect the native flora within patches, ecological restorative efforts should be applied to matrix lands by creating better connectivity.

### *Importance of Biodiversity: Indian Status*

Today merely 10 to 20 plant species provide 80-90 per cent food requirements of the world. In India, rural communities, particularly the tribals obtain considerable part of their daily food from wild plants. At one time, nearly all medicines were derived from biological resources. Around 20,000 plant species are believed to be used for medicine in the developing world. In India the knowledge about medicinal value of plants has evolved in the form of traditional systems of medicinal sciences, like Unani, Ayurveda and Siddha. More than 8,000 species are used in some 10,000 drugs formulations. The source of 90 per cent of industry's requirement is from forests. It is estimated that about 0.5 million ton (dry weight) of plant material is collected each year from the forests and often the method of collection is destructive. The global plant based drug trade is projected around Rs. 310 billion with a 7 per cent annual growth rate but India has only 2.5 per cent share in it. The outstripping supply has put unreasonable pressure on our wild phyto-resources. Due to this, certain species are at risk and others will probably join. Biotechnology provides useful tools in the assessment and monitoring of biodiversity both *in-situ* and *ex-situ* situations.

**Biodiversity Assessment**

In India, formal policies and programmes for conservation and sustainable utilisation of biodiversity resources date back to several decades. The concepts of environmental protection are enshrined in the Indian Constitution in Articles 48a and 51a(g). Major Central Acts relevant to biodiversity are—Forest Act, 1972, Wildlife (Protection) Act, 1972, Forest (Conservation) Act, 1980 and Environment (Protection) Act, 1986. The various Central Acts are supported by a number of state laws and statutes concerning forests and other natural resources. Policies and strategies directly relevant to biodiversity include National Forest Policy amended in 1988; National Conservation Strategy and Policy Statement for Environment and Sustainable Development; National Land Use Policy; National Policy and Action Strategy on Biodiversity; National Wildlife Action Plan and Environmental Action Plan.

Adopting a consultative process with the stakeholders, a National Policy and Action Strategy on Biological Diversity has been drawn up as a macro level statement of strategies. This plan aims at consolidating the on-going efforts of conservation and sustainable use of biological diversity, identifying gaps in various sectors and providing a policy and programme regime to ensure attainment of the three objectives of the CBD. Likewise Ministry of Environment and Forests (Government of India) has recently launched project to document country's biodiversity and evolve action plan to protect country's biodiversity—National Biodiversity Action Plan. This framework document provides the basis for developing detailed action at micro and regional levels. Identifying critically important areas and gaps in taxonomic work, an All India Coordinated Project on Taxonomy (AICOPTAX) has been launched for prioritised purpose oriented capacity building in taxonomy.

The diversity of the country's biological resources is yet to be fully known. Approximately, 65 per cent of the total geographical area has been surveyed so far. Based on this, over 46,000 species of plants and 81,000 species of animals have been described by the Botanical Survey of India (BSI) established in 1890 and Zoological Survey of India (ZSI) established in 1916 respectively. The Forest Survey of India established in 1981 assesses the forest cover, with a view to develop an accurate database for planning and monitoring purpose. According to an estimate (BSI, 1983) about 30 per cent plant species are endemic to India. Areas rich in endemism are also reported from Eastern Ghats (MacKinnon and MacKinnon, 1986). The Project on Study, Survey and Conservation of Endangered Plants (POSSCEP) has estimated that about 3000-4000 plant species are under different degrees of threat. The red data book enlists the IUCN categories of plants and animals occurring in various parts of the nation (Nayar and Shastri, 1987).

## Measuring Biodiversity

In recent years we are confronted with a situation wherein we need to know the methods to measure biodiversity, so that we can determine where *in-situ* conservation action is required; rather than how to implement the site-specific biodiversity conservation plans. The latter is ground-based and enough scientific options are available. The measure of biodiversity should quantify a value shared broadly, the people for whom they are acting. Arguments for measuring biodiversity value as character *richness* do at least provides a reasonable starting point, as one possible answer to the question of what is valued in diversity. The remote predictors or surrogates often play very significant role to measure richness. The habitat surrogates including classification of vegetation, details on the physical environment, factors determining the biodiversity loss in a spatial context may be of practical information value and could reduce sampling intensity. This information base could also guide detailed sampling on the ground. These large scale surrogates include entire functional system and are more likely to promote population viability in the ecosystem. If the value of biodiversity to a conservationist is associated with its use to people then this ought to be separated carefully from issues of rarity, viability and threat. If the biodiversity value is associated with richness in a currency of characters of organism then the higher level of biological organization (or environmental factors affecting its distribution) will have to be used as surrogate measures (Fig. 1).

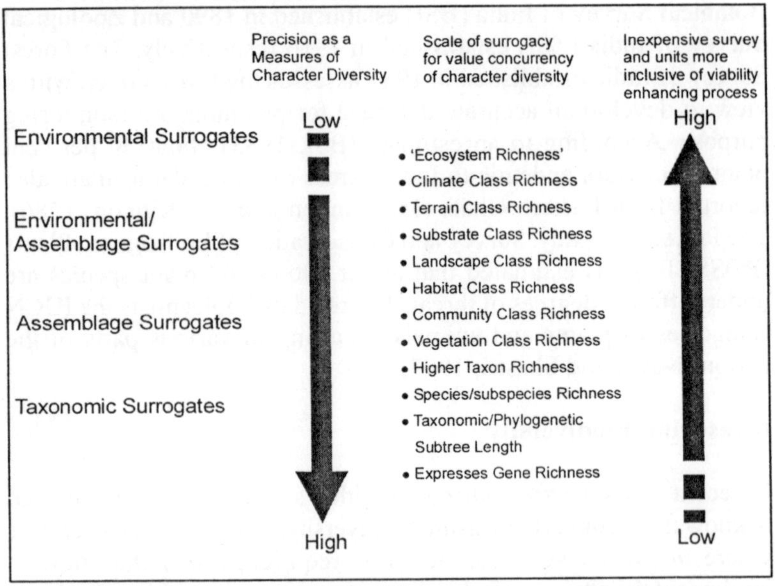


**Fig. 1:** Surrogates in Measuring Biodiversity Value (Anon., 2002 a, b, c).

### *Landscape Characterisation*

Human activity has widespread impact on biodiversity, affecting ecological entities from species to whole communities and ecosystems, though heterogeneity in the landscape can be due to moderate disturbance. Disturbance regimes and their impact on communities and landscape can be well understood by analyzing spatial and temporal architecture of disturbance (Moloney and Levin, 1996).

Landscape ecology examines spatial variation in fragmentation at a variety of scales. It includes the biophysical and societal causes and consequences of landscape heterogeneity. Landscape ecology embraces spatial heterogeneity and pattern: how to characterize pattern, how it develops and changes through time, and its implications for populations, communities, and ecosystem processes. Because landscapes are large they are typically measured in kilometres. Human influences are an important influence on landscape pattern and landscape ecology. Hence, landscape ecology quite naturally is concerned with anthropogenic aspects of landscape pattern and change.

Integration of regional vegetation, physiography and climate provide details on landscape characteristics. The empirical assessments of the patterns of land use and field studies explain the processes behind land use and land cover change over varying spatial and temporal scales. These global and national level attempts have immense value in understanding the patterns and processes of land cover and land use change, human responses to land use/land cover change and development of databases on land surface, biophysical processes and their drivers. Land use/land cover change is central to the interests of the science of global environmental change (climate change, loss of biodiversity and the sustainability of human-environment interactions).

To understand recent changes in the Earth living system, it needs quantitative, spatially-explicit data on how land cover has been transformed by human use over the last 300 years and how it will be changed in the next 50-100 years. Such data needs to be generated by use of space potentials to derive required data sets to study the vegetation cover and related parameters. To generate scenario on future modifications of the Earth system, one has to project how changes in land use are likely to affect land cover in the future using integrated land use change models. Measurements of past rates and spatial patterns of land cover changes can be derived for the recent past (last 30 years). Models of land use at regional to global scales provide valuable inferences on economy-environment linkages. They have to cope with such issues as heterogeneity and scales, technological innovations, policy and institutional changes, and rural-urban dynamics.

**Geoinformatics**

Ever-growing understanding and acceptance that the Earth functions as a complex system composed of myriad interrelated mechanisms have made scientists realize that existing information systems and techniques used are inadequate. Currently, the uncoordinated distribution of available data sets, a lack of documentation about them, and the lack of easy-to-use access tools and computer codes are major obstacles for scientists and educators alike. These obstacles have hindered scientists and educators in the access and full use of available data and information, and hence have limited scientific productivity and the quality of education. Recent technological advances, however, provide practical means to overcome such problems. Advances in computer design, software, disk storage systems as well as the growth of the World Wide Web (www) now permit for the first time the management of gigabytes to terabytes of data for distribution to scientists, educators, students, and the general public. Geoinformatics (geographic information science, geomatics) aims at the development and application of methods for solving specific problems with special emphasis on the geographical position of objects.

The environmental control on the forest vegetation is well documented (Muller-Dembois and Ellenberg, 1974). Physiography, topography, climate and human interventions largely control the distribution of vegetation and biodiversity. The developments in computer based Geographic Information System (GIS) enables the integration of spatial and non-spatial information for defining the habitats and improving vegetation type descriptions in space and time. A review of GIS and databases for vegetation mapping and monitoring is given by Skole *et al.* (1993). It is also possible to evolve geospatial models using multi-criterion to present disturbance regimes and landscape diversity. Landscape ecology has evolved as an operational tool with the availability of geospatial modelling techniques. Antenucci *et al.* (1991), Miller (1994) and Tomlin (1990) provide number of examples of application relevant to biodiversity. The power of having all information and knowledge along with access, modelling, and visualization tools at the finger tips of a user has great potential in advancing science, accelerating the discovery process, and enhancing the quality of science and education. The step in Geoinformatics for creating Biodiversity Information System is shown in the Fig. 2.

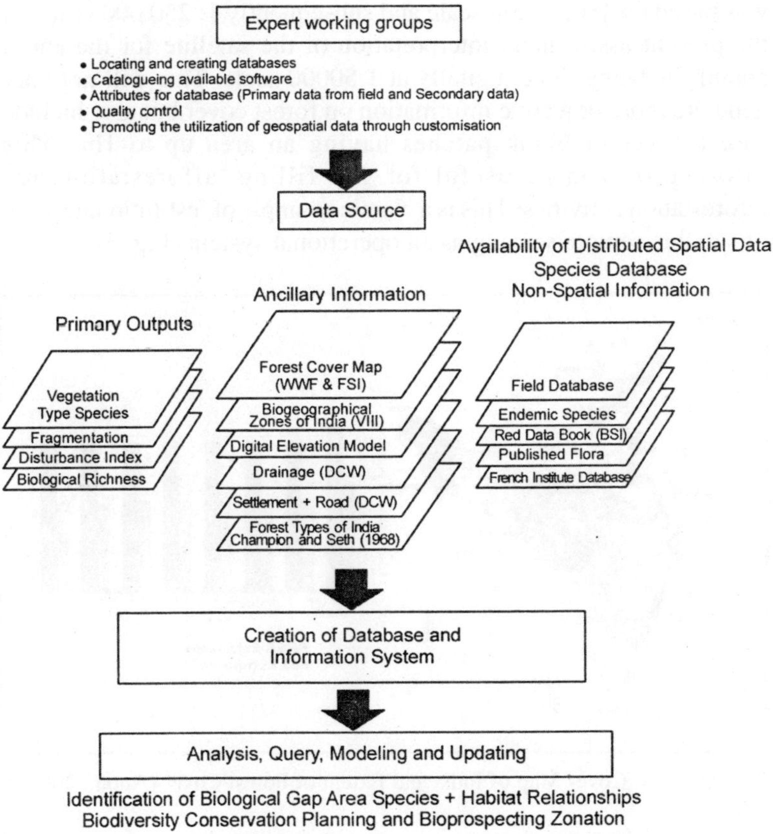


**Fig. 2:** Biodiversity Information System using Geospatial Tools (Anon., 2002 a, b, c)

## National Forest Cover Assessment

Assessment of the forest cover of the country, based on satellite data interpretation is one of the important initiatives of National Remote Sensing Agency (Department of Space) during 1983. For the first time country's forest cover assessment carried out using Landsat-MSS false colour data sets of 1972-75 and 1980-1983 periods to categories forest cover into different density classes on 1:1 million scale. Forest Survey of India (FSI) subsequently, took over the task of providing on the biennially basis. The first assessment in 1987 was based on 1:1 million scale and subsequently 1: 250 000 scale. In the present assessment, interpretation of the satellite for the entire country is being done digitally at 1:50000 scale. This will not only generate more objective information on forest cover but also include forest cover or blank patches having an area up to 1ha. This information can be useful for gap filling, afforestation and reforestation activities. This is a classic example of institutionalisation of satellite remote sensing as an operational system (Fig. 3).

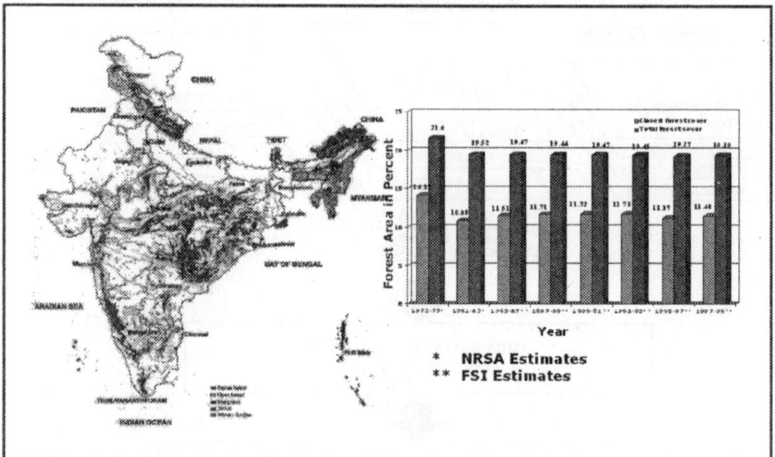


**Fig. 3:** Forest Cover Map of India and Extent of Forest Cover (Anon., 2002d).

### *Biodiversity Characterization at Landscape Level*

Biodiversity characterization at landscape level is attempted in biodiversity hotspots (Anon., 2002a, b, c) and now Central India and Eastern Ghats of the country using satellite remote sensing and GIS. The study is particularly important from scientific point of view as large diverse landscapes have been analysed at spatial extent and quite significant results on fragmentation, disturbance regimes and biological richness. Findings revealed that high fragmented forests have fewer plant species, anthropogenic use evidence and newer community types than medium and low fragment forests. It may also be concluded that slow fragmentation produces a decreasing trend in plant species diversity as revealed from the study. The landscape perspective is fundamental to understanding the role of disturbance.

Disturbance regimes vary across landscapes as a function of topography and substrate (Fig. 4).

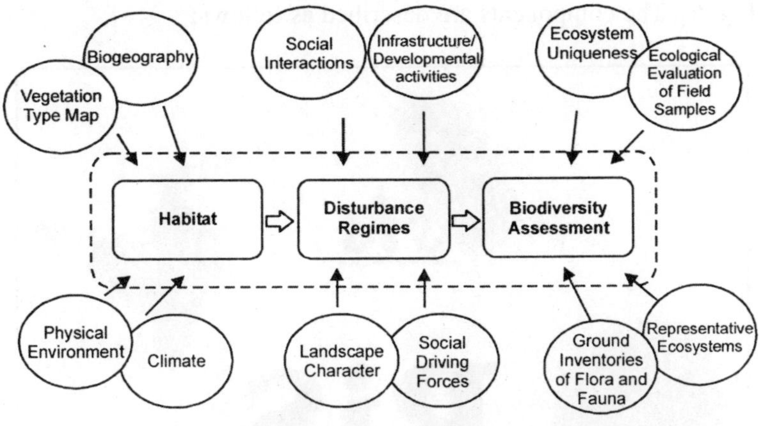


**Fig. 4:** Approach for the Biodiversity Assessment (Roy and Tomar, 2000—revised).

Many of the natural ecosystems have become smaller and their surroundings have been modified. The biological richness is related to higher species diversity, higher degree of ecosystem uniqueness, higher degree of terrain complexity and lower levels of disturbance. Another dimension, socio-economic value of each species was also considered for biological richness. The degraded and barren areas were found to least biologically rich. The biological richness maps very well reflect the distribution and status of biodiversity in the region. The information generated in this study would be of immense use if linked to other planning and policy-making that directs resource allocation, proper land use planning and sustainable consumption of natural resources.

### *Biodiversity Information System (BIS)*

The Department of Biotechnology (DBT) and Department of Space (DOS) have joined hands to undertake study on biodiversity characterization. The main objectives of this project are to identify the biorich areas, to evaluate the forest types for their value and to provide locational information of the economically important species for bioprospecting. The outcome of this project, Biodiversity

Information System (BIS) targets to provide user level information regarding the Biodiversity hotspots (Anon., 2002d). The Biodiversity Information System was conceptualized as a 5-component unit (Fig. 5). The components are described as follows:

**Fig. 5:** Components of Biodiversity Information System (Anon., 2002e).

*BioSPATIAL (Biodiversity characterization at Landscape Level)*

The BioSPATIAL Query Shell is a spatial component of the Biodiversity Information System. In this, the client can view maps from the regional level to the state level. The current studies in Biodiversity characterization cover four major biodiversity hotspots of the country. The query shell starts from regional level and zooms down to the local (state) level. This web enabled interface will immensely help in monitoring biodiversity and landscape changes in these hotspots and would thus help in the conservation of the same.

*BIOSPEC (Bioprospecting and Molecular Taxonomy Programme)*

The Department of Biotechnology has a major project titled the Bioprospecting and Molecular Taxonomy Programme. Under this project rare and important species, which include names like *Podophyllum hexandrum* ex Royle, *Gaultheria fragrantissima* Wall etc. are being studied. This query shell also integrates the data from the published books from the Botanical Survey of India. The species, which have been located on the maps, have been classified into eleven categories as per the IUCN standard. This system correlates the species with the Biodiversity Characterization information.

*FRIS (Forest Resource Information System)*

Forests are an important resource of our country. Forests have acquired increasing importance in the recent past for their role not only in meeting the material requirements but also for their ecological and environmental functions. Authentic historic data on recorded forest area is not available. Therefore, accurate and updated information on forest resources is of paramount importance. Some of the important parameters used for measuring forest resources are forest cover, growing stock, annual increment, species composition, biodiversity, non-timber forest products etc. This site deals with providing the same information to Environmentalists, Researchers, Planners, Administrators, Decision-makers and Local Communities to provide them with sound, adequate and updated information on forests of India and on role of Remote Sensing and Geographic Information System, thus supporting a holistic approach to Planning and Forest Management.

*PhytoSIS (Species Information System)*

A large ground sampling exercise, which had been conducted to characterize vegetation type, species richness, ecosystem uniqueness etc. has been collated. The species information system enables the client to perform direct and indirect queries on species. For example, a user may at one hand choose the name of the species for details, and at the other hand, select certain parameters which result in a list of species.

*BioconSDSS (Spatial Decision Support System)*

The maps which have been generated as a result of large scale and in depth research are the outputs from customized software generated by the Department of Space, called "Biodiversity Characterization Program (Bio_CAP)". However, one can question whether the client can generate map by varying the parameters in the study and changing the weightages as considered in the case of Biodiversity Characterization. An effort to generate the web enabled interface for the spatial decision support system has been made. The facilitated the user to change the parameters as per site specific requirement or objective of particular research.

***GAP Analysis for Conservation Prioritisation***

One of the most important component of biodiversity conservation is to maintain continuity of natural landscapes, vital corridors and natural plant community settings. This approach will provide habitats to the associated life forms—a basic requirement of *in-situ* conservation. The visualization of natural landscape model helps to simulate the association. Satellite image derived information provide spatio-temporal scenario and helps in simulating natural settings. The conservation of the diversity require specific efforts to restore the gaps created by natural or anthropogenic process. Overlaying of habitat maps, biological richness and existing protected area network provides gaps in landscape conservation (Fig. 6). Such information is vital to policy-makers and planners in developing network of protected area representing all habitats.

## Discussion

The loss of natural habitats and their isolation is major component of eroding genetic resource base in the biosphere. The conservation priority setting varies considerably due to complexity of biodiversity and the number of ways of valuing it. Ecological approaches for setting priorities for biodiversity conservation generally seek to protect most of the species within conservation areas that are representative of a region's natural habitat. Ecosystem approaches for identifying conservation priorities using biological criteria. Among the biological criteria are: *richness* (the number of species or ecosystems in given

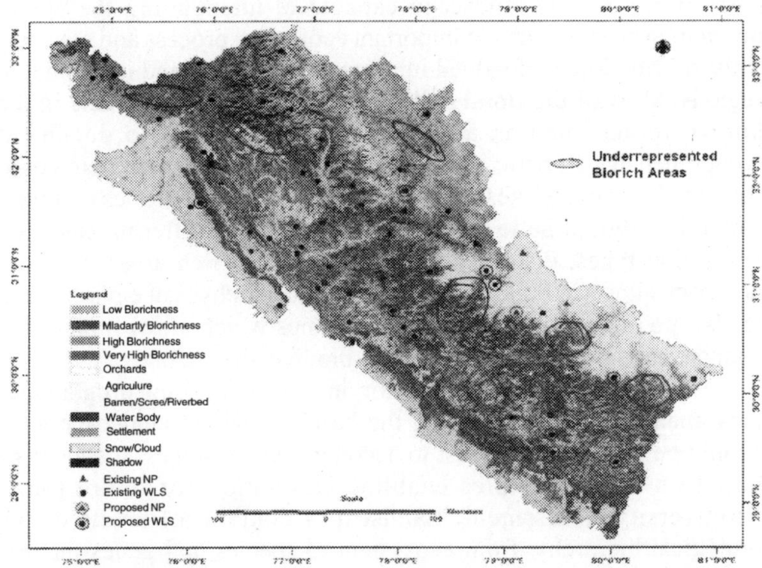


**Fig. 6:** Underrepresented Biorich Areas in the Protected Area Network of Western Himalayan Region of India (Chandershekhar *et al.*, 2002).

area), *rarity*, *threat* (degree of harm or danger), *distinctiveness* (how much a species differs from its nearest relative), *representativeness* (how closely an area represents a defined ecosystem) and *function* (the degree to which a species or ecosystem affects the ability of other species or ecosystems to persist). It also considers the physical environment, ecological processes and disturbance regimes that help to define the ecosystems (Roy and Tomar, 2000). Some priority setting approaches use social, policy and institutional criteria as well. *Utility*, the most common non-biological criteria, points to biodiversity elements of known or potential use to humankind. *Feasibility* is often considered paramount in deciding how to allocate conservation resources. Geospatial techniques can suggest the actions that are most likely to succeed taking into account social, political and institutional framework.

The assessment of biological rich areas using aerospace data develops better understanding of the landscapes as driven by pattern of richness, endemism, biological corridors, community, composition and diversity. The analysis made also presents full range of distinct natural communities and ecological status at landscape level. Satellite maps bring out the landscape capable of maintaining the viable population species; sustain important ecological process and services. This information is of valued importance in rugged and inhospitable regions. Most of the floral collections in the region are made in the known route. Such an approach has not allowed to document comprehensive floristic inventory. The results presented here could form the basic guideline to plan floral and faunal inventories in future. The focus should be to cover varied landscapes differing based on vegetation types, disturbance regimes and bio-rich areas. Such an approach allows to build habitat factors like, biophysical environment, landscape indices and disturbance regimes which allow monitoring changes taking place over a time in biodiversity regimes.

The goals and scales of inventorying and monitoring programmes may change with time. Hence, the baseline data at landscape level should be sufficiently robust to accommodate changes. It should be based on robust samples enabling calibration for future Rapid Biodiversity Assessment. Landscapes contain all levels of the biological hierarchy, from ecosystem to species and genes that are targeted for biodiversity inventories and conservation. The DOS-DBT effort to characterize vegetation cover, fragmentation, disturbance and biological richness across the landscape is an unique initiative at national level. The database is also organized in the form Biodiversity Information System (BIS). The species database has been linked with above spatial details. The BIS allows to identify gap areas, species/ habitat relationship and helps in biodiversity conservation planning by setting priority areas. Such database coupled with detailed site-specific field inventories provide information on bio-prospecting.

The baseline information generated in the DOS-DBT project has relevance to the international program of the biodiversity conservation and management—DIVERSITA's (an international initiative on the biodiversity assessment). The issues focused by DIVERSITA's in core project Global Invasive Species programmes (GISP) and the Global Mountain Biodiversity Assessment (GMBA) coincides with the Indian initiatives addressed by DOS-DBT project. The biological diversity is considered essential for the persistent functioning and integrity of mountain ecosystem and this dependency likely to increase as environmental conditions change. Steep terrain and mountain climate in combination with ever increasing land use pressure cause mountain ecosystems to rank among the most endangered landscape worldwide.

The core project 1 of DIVERSITA's for the understanding, monitoring and predicting biodiversity changes looks into (i) assessing current biodiversity; (ii) monitoring biodiversity changes and (iii) understanding and predicting biodiversity changes. The landscape approach for the characterization of biodiversity using remote sensing and GIS has carried out the aforesaid objective in the hot spots of the Indian subcontinent. The core project 2 views the Assessing impacts of biodiversity change for assessing (i) impacts of biodiversity changes on ecosystem functioning and ecosystem services; (ii) impacts of biodiversity on human and livestock health. The national initiative of the DOS-DBT has partially touched these issues and is in process of acknowledging the remaining.

Finally, geospatial tools provide vantage platform to record these changes in space and time. The varied satellite systems allow generating information base at global, regional and local level. Space science has developed indigenous state of art platforms with continuity of data products. The landscape ecological concepts help us to understand the biodiversity distribution in the region in varying spatial scales.

## REFERENCES


Anonymous, (2002): Biodiversity Characterisation at Landscape Level in North-east India using Satellite Remote Sensing and Geographical Information System. Indian Institute of Remote Sensing (NRSA), Dehradun, p. 296.

Anonymous, (2002b): Biodiversity Characterisation at Landscape Level in Western Ghats India using Satellite Remote Sensing and Geographical Information System. Indian Institute of Remote Sensing (NRSA), Dehradun, p. 244.

Anonymous, (2002c): Biodiversity Characterisation at Landscape Level in Western Himalayas India using Satellite Remote Sensing and Geographical Information System. Indian Institute of Remote Sensing (NRSA), Dehradun, p. 234.

Anonymous, (2002d): Annual Forest Report 2001-2002. *Forest Survey of India, MEoF*, Government of India, Kaulagarh Road. Derhadun, p. 29.

Anonymous, (2002e): *Biodiversity Information System*. Indian Institute of Remote Sensing (NRSA), Dehradun, p. 94.

Antenucci, J.C., Brown, K., Croswell, P.L. and Kevany, M. (1991): *Geographic Information System: A Guide to the Technology*, Van Nostrand Reinhold, New York.

BSI, (1983): *Flora and Vegetation of India—An outline*. Botanical Survey of India, Howrah, pp. 2331-2346.

Chandrashekhar M.B., P.S. Roy and Sarnam, Singh. (2002): Assessment of underrepresented biorich areas in the protected areas network (PAN) of Western Himalayan Region: GAP based approach. *Asian Journal of Geoinfromatics*. (Communicated).

Dale, V.H., S.M. Pearson, H.L. Offerman, and R.V. O'Neill. (1994): Relating patterns of land-use change to faunal biodiversity in the Central Amazon. *Conservation Biology* **8** (4): 1027-1036.

Gardner, R.H., R.V. O'Neill, M.G. Turner and V.H. Dale. (1989): Quantifying scale—dependent effects of animal movement with simple percolation models. *Landscape Ecology*, **1**: 19-28.

Gilpin, M.E. and M.E. Soule. (1986): Minimum viable populations: processes of species extinctions. In: M.E. Soule (ed.). Conservation biology: the science of scarcity and diversity, pp. 19-34. Sunderland, Mass: Sinauer Associates, Inc.

Holl, K.D. (1999): Factors limiting tropical rain forest regeneration in abandoned pasture: seed rain, seed germination, microclimate, and soil. *Biotropia* **31** (2): 229-242.

Mayers, N. (1992): *The Primary Source: Tropical Forests and Our Future*. New York: Norton.

Miller, R.I. (1994): Mapping the Diversity of Nature. Chapman and Hall, London.

Moloney, K.A. and S.A. Levin. (1996): The effect of disturbance architecture on Landscape level population dynamics. *Ecology*, **77**: 375-394.

Muller-Dembois, Dietor and Ellenberg, D.H. (1994): *Aims and Methods of Vegetation Ecology*. John Wiley & Sons, New York.

Nayar, M.P. and Shastry, A.R.K. (eds.) (1987): *Red data Book of Indian Plants*, Vol. 1. Botanical Survey of India. Calcutta, pp. 367.

O'Neill, R.V., Krummel, J.R. and Gardner, R.H. (1988a): Indices of landscape pattern. *Landscape Ecol.***1**: 153-162

O'Neill. R.V., B.T. Milne, M.G. Turner and R.H. Gardner. (1988b): Resource utilization scales and landscpe pattern. *Landscape Ecology*, **2**: 63-91.

Pimm, S.L., G.J. Russel, J.L. Gittleman and T.M. Brooks. (1995): The future of biodiversity. *Science* **269**: 347-350.

Prance, Ghillean T. (1982): Forest Refuges: Evidence from woody angiosperms, pp. 137-157. *In*: G.T. Prance (ed.) *Biological Diversification in the Tropics*, Columbia University Press, New York.

Roy, P.S. and Sanjay Tomar, (2000): Biodiversity Characterisation at Landscape level using Geospatial Modelling Technique. *Biological Conservation* **95**(1): 95-109.

Roy, P.S., P.K. Joshi, Gautam Talukdar, M.B. Chandrashekhar, C. Jeganathan, Sameer Saran, S.P.S. Kushwaha, Sarnam Singh, R.S. Tripathi, M.C. Porwal, A.K. Tiwari, S. Sudhakar, Harnam Singh, T.P. Singh, Uday Kumar and Sanjay Tomar. (2002): Biodiversity characterization at Landscape level and defining conservation strategies. Forest Cover Assessment in Asia: *Proceedings of the International Workshop on Asian Forest Cover Assessment & Conservation Issues, 12-14 February, Dehradun, India*, pp. 193-206.

Roy, P.S. (2003): Biodiversity Conservation: Perspective from Space (Unpublished work).

Saunders, D.A., H.J. Hobbs, C.R. Margules. (1990): Biological consequences of Ecosystem Fragmentation: a review. *Conservation Biology* **5** (1): 18-27.

Skole, D.L., Moore, B. and Chomentowski, (1993): Global Geographic Information Systems. In: Solomon, A.M. ansd Shrugart, H.H. (eds.), *Vegetation Dynamics and Global Change*. Chapman and Hall, New York, pp. 168-189.

Tomlin, C.D. (1990): *Geographic Information System and Cartographic Modelling*. Prentice-Hall, New York.

Turner, I.M. and Corlett, R.T. (1996): The conservation value of small, isolated fragments of lowland tropical rain forest. *Trends in Ecology and Evolution* **11**: 330-33.

UNEP, (1995): *Global Biodiversity Assessment*. Cambridge University Press, New York.

Whitmore, T.C. (1997): Tropical forest disturbance, disappearance, and species loss, pp. 2-28. In: Laurence W.F. and Bierregaard, Jr., R.O. (Editors), *Tropical Forest Remnants*. The University of Chicago Press, Chicago.

Williams-Linera, G, V. Domiguez-Gastelu, and M.E. Garcia-Zurita. (1998): Microenvironment and floristics of different edges in a fragmented tropical rainforest. *Conservation Biology* **12** (5): 1091-1102.

# 4

# Land use Interpretation from Satellite Imagery

## *Geomorphic Controls on Land Use and Settlement in Narmada Valley and Adjoining Hills and Plateaus*

J.L. JAIN

### Introduction

Land use and settlement pattern of an area constitutes essence of its geographic personality and lends it a distinctive character. It also marks a true interface of physical and the human/cultural aspects of the local/regional geography and as well reflects the cumulative effects of its historical development or sequent occupance. The present study delves into these aspects of geographically the very core region of India, as they are discernible through the visual analysis of satellite imagery and basic maps of the region. It first considers rather old data situation—in the early seventies, and compares, some important parts with the situation obtaining at the end of the century. One important aspect of the study is that it correlates land use with the geologic/geomorphic factors and presents a model of land use pattern much as a geological section across the structural grain of the region.

### The Region and its Geographical Framework

Landsat I MSS Scene No. 155-044 for the February 1973, on which this study is based, covers a large (i.e., well over thirty thousand square km) Central part of erstwhile Madhya Pradesh State (i.e. including the Present Chhattisgarh State also), encompassing Jabalpur, Damoh, Sagar, Narsimhpur and Seoni districts, and sizeable parts of Mandla, Hoshangabad and Chhindwara districts. It stretches from 22.5° North to 24.5° North latitude and from 78.5° East to about 80.3° East longitudes. Changes in land use and settlement pattern for important areas were noted with help of IRS 1-B and 1-D Geocoded data.

Structurally, the area is predominated by the great Narmada trough running nearly East-West in the Central part (Fig. 1+4). It is flanked to the South by the complex hills and ranges of the Satpura Range made largely of Achaean and granitic rocks. The Northern rim of the Narmada Trough is more clearly and sharply defined by the Southern scarp face of the Bundelkhand Plateaus and marked by Bhanrer Range and farther East by the Kaimur Range, which themselves are all part of the great Vindhyachala Range. This range and the undulating plateaus to the North are essentially a Vindhyan sandstone/shale rock formation, overlain by basaltic traps that become thicker in the North and West.

In the Western part these ranges are less sharp and less continuous, where they simply form the scarp face of the basaltic Malwa Plateau. Farther East, these ranges veer to the North-east, leaving a broad triangular plain between the Narmada River and these scarp ranges. Intervening the Northern plateaus are several lesser hill-ranges trending North or North North-East and exposing the sandstone hills from below the more easily weathered and eroded basaltic traps. These plateaus are drained to the North or North-East by the Sonar-Bearma river systems.

The Narmada Basin is essentially a structural trough or rift between the Satpura and the Vindhyachala Ranges. Below the thin veneer of alluvium, the basin floor is made up of Achaean rocks with prominent dykes cropping out at places. The mighty Narmada River draining the basin to the West, keeps its course closer to the Northern scarp in this part. Narmada receives fewer tributaries from the North, and many more and bigger ones from the South, which are also broader and quite full of alluvial detritus. Nevertheless, River Hiran in the Eastern part and Singhore and Tendoni-Barna river systems in the Western part of the basin are prominent Northern tributaries of the Narmada, which have each distinct geomorphic and land use characteristics. However, the upper course of Narmada, i.e. East and South of Jabalpur, is actually outside the structural trough, which forms Western part of the highly undulating Maikal plateau.

## Land use Categories and their Distribution

On the basis of Green and Red bands imagery B/W prints and a colour composite of the Green, Red and NIR bands as standard FCC, and relevant maps as well local knowledge and/or circumstantial evidences, following categories of land use/land cover are easily discernible. These classes are further divisible into sub-classes in some cases, whereas there are composite or mixed classes and/or sub-classes apparent in some areas. These classes are: Forest, Cropland, Scrub and Uncultivable Wasteland, Cultivable Wasteland, larger Built-up areas, and Water Bodies[1]. Scrub and Uncultivable Wasteland are liable to be confused, which can be further confounded with much degraded Forest patches/pockets. Depending on the type of crop and the season of the data, Cultivable Wasteland, Fallow land and Cropland are again liable to confusion.

### *Forest*

Much of the area under study is forested, which in this part of the country are of Tropical Wet Deciduous and Tropical Dry Deciduous[2] type, with Sal and Teak as the principal timer species. Their actual composition is, though, significantly variable—depending on physiography, aspect, moisture, successional status, and human intervention/exploitation. On the very coarse data of Landsat MSS there appears to be sizeable area as mixed or transitional between forests, cultivable wastes and fellow lands—as over extensive South-Eastern and the North-Western tracts; and scrubland and cultivable wasteland.

Extensive patches of good/healthy forests are to be seen (Fig. 2)[3] in: (A) The Southern and Eastern part of region—mostly on the Satpura highlands (Wet Deciduous). (B) A rather narrow belt covering higher reaches of the Vindhyachala Hills and ranges, running essentially East-west across the Central part of the region (often Wet Deciduous). Certain forested hills and ranges of the Bhanrer Range in the Eastern part stand out sharply and enhanced due to aspect, sun-angle, and long thin shadows associated with them. Further North-East this forest belt broadens out Northwards, though in a very fragmented manner. (C) In the extreme South-western part lies the most extensive and richest forest patch on Pachmarhi Hills of the Satpura Range. Within that patch, however, there is an extensive basin-like patch with much cultivable wastes and fallows.

Of the numerous forest patches only few survive with good health and vigour, and very many others are degraded open forest or scrubland and easily confused/mixed with cultivable or uncultivable wasteland. Such patches are particularly seen in the Satpura highland as well as Vindhyachala ranges and the latter's several tentacles extending to the North. Many of the areas are recognized as mixed land uses, such as forests and cultivable wastes and fallows in the South-eastern (Eastern Satpura and upper Wenganga) and the North-Western part (Sagar plateau), where the very small-small patches of cropland, cultivable wastes, and scrubs are very intricately inter-mixed, making delineation/regionalisation extremely tedious at this scale. In the extreme North-east the rock formation changes to Bijawar meta-sedimentary which add a light purple over-tone to other hues.

### *Cropland*

As a category of vegetation, cropland, particularly standing crops, share much with forests with respects the spectral signature, more so in late January in Central India when the deciduous forests have luxuriant foliage and rabi crops like wheat are still lush green. However, croplands do present some fine texture and a mat-finish rather a glossy appearance of rich, dense forests. They may again be diverse finer variations according as the stage and type of crops, their luxuriance, and backdrop of soil. So too, croplands recently harvested may be compounded with fallows or cultivable wastes, i.e. predominantly rabi or kharif nature of crops grown is likely to have a dramatic bearing on the spectral signature of the croplands. In the late winter season, when this data was acquired, this part of the country has kharif crops harvested already, whereas rabi crops are at full growth and luxuriance. Therefore areas predominantly given to kharif crops may look very lean or even fallows.

In this area of heterogeneous topography and lithology, distribution of cropland is also much varigated. However, there are three extensive areas of standing rabi crops (mostly wheat). Narmada Basin in general is the principal crop region of the Scene. In the upper Narmada Basin, the triangular interfluves between Narmada and its principal right-hand tributary Hiran River comprises the most conspicuous and richest cropland, which extends eastward amid forested hills. In this Haveli plain the river Hiran first courses North and then curves sharply to the South-West washing the foothills of the Bhanrer range for a long distance before pouring into the Narmada close to the Northern Wall of the Narmada Basin/Trough. Numerous lakes and ponds of all sizes, which continue to occur in and around Jabalpur city also, further enrich this plain.

Farther west—in the middle stretch of Narmada—there are many tributaries and sub-tributaries from the South enriching the broad Southern flank of the basin with fresh alluvium and groundwater. Closer vicinities of the trunk streams as well as the tributaries, however, appear devoid of crops, i.e. as sandy and/or revinous cultivable wastes and fallows. One should beware a stumbling block here—much of this fallow-looking land may actually be a kharif cropland (mostly sugarcane and rice), which have been harvested by this time. In the Western-most stretch of the Basin, Narmada and its many tributaries are more revinous and many also have broad sandy wastes. The cropland here is more scant and patchy, leaving notable belts of revinous and/or sandy wastes in between croplands father a field. This whole area also shows a purple overtone of hue due to black soil of the basaltic top cover found particularly in the Western part of the Scene.

Another notable cropland in the broad eastern flank of Sonar Valley—more precisely between the Sonar and its right-hand tributary, Kopra river. Extent of cropland on Western flank of the Sonar is comparatively smaller, poorer and rather broken. Bearma river occupies an extensive plateau which is largely rocky with rather skeletal soils. Croplands here are confined to the close vicinity of the middle stretch of the Bearma. Father afield, particularly in upper Bearma basin, the area essentially rocky waste. Also called Damoh Plateau, the land is here more rugged and drier. It distinctly betrays numerous fine fractures amid the rocky wastes as fine red lines, as they hold better soil moisture and support plant growth. Junctions of two or more such fractures result in small specks of cropland and a settlement nearby. Farther East this country merges with the complex hills covered mostly with forests or stony wastes. West of the Sonar river, Sagar plateau shows numerous tiny patches of moderate croplands interspersed with other types of land use.

### *Scrubland and Uncultivable Wasteland*

Though they form two separate categories of land use, they are treated here together as they show some common associations with terrain character. Scrubs merge through degraded forests with standing forests, and appear to be associated with patches of hard, poor skeletal soils with low moisture, and undulating broken surfaces. Uncultivable wastes comprise barren lands such as bare rock outcrops, steep and/or dissected slopes, too stony to support any appreciable amount of plant growth. Spectrally, there is but a subtle difference between the appearance of the two categories—scrubs appearing light brown, and the wastes appearing in the white to brown grayish hues. Scrubs also show comparatively rough texture and irregular pattern.

Scrubs predominate in the North-eastern part of the Scene, where it interspersed extensively within uncultivable wastes. One more notable patch of scrubs occurs amid Satpura Hill in the South-western part of the Scene interspersed with small open forest patches. Uncultivable wastes like dry, stony, rugged land occur extensively in the upper Bearma basin, more so in the right-hand flank of the basin. Farther North-east they get merged and intermingled with scrubs and degraded forest patches around cores of good forest on hills. A very different type of wasteland occurs as sandy and/or revinous belts along some rivers in the Western of Narmada Trough. Some stony wastes in and around Jabalpur are likely to compound with built-up areas, accept for their irregular shapes.

### *Cultivable Wasteland*

This category is treated here as inclusive of fallows also. Any land which can be cultivated or which has been cultivated sometimes and which is currently out of cultivation comes under this category. Being close to settled areas and cropland as well as sharing much spectrally with any bare ground, cultivable wastelands are much amenable in interpretation to be compounded with harvested fields. Therefore, other background or circumstantial corroboration is needed/desirable. Reflecting highly in the Green and Red bands and poorly in the NIR band, this type land use manifests in light cyan hue.

Cultivable wastes and fallow are often found interspersed with other land use types. Nevertheless this category is found more frequently and larger patches in middle Narmada Trough close to the many rivers that are revinous and/or sand, sometimes broad also. Another long and conspicuous patch of this land is seen on the Eastern edge of Sagar Plateau, for reasons of the Lameta rock formation.

***Built-up Areas***

Only larger urban places are readily detectable on this small-scale data of low resolution, while moderate and smaller towns can often be deduced. Sizeable urban built-up areas appear bluish-green mixed with whitish gray and red dot of lawns, parks, etc. Routes also appear in similar but brighter hues. Jabalpur city near the eastern margin is unmistakable. Medium cities like Sagar can also be seen easily, and even towns like Damoh can be picked up on their own. Smaller places on the data of this scale and resolution are often lost in the spectrally similar surroundings like wastes and fallows. That is why sizeable towns, e.g. Narsimhpur and Lakhnadon appear indistinct and insipient. However urban areas do show some texture and some pattern of roads and blocks. Smaller places can be deduced with additional elements like convergence of routes, which themselves may or may not be readily visible on the coarse resolution, small scale data.

***Water Bodies***

River, canals, lakes-ponds, tanks and reservoirs are the only water bodies in this mid-land scene. Though a body of water by itself should appear very dark with little bluish tinge, its actual look is compounded by reflection from shallow bottoms, suspended particulate material (turbidity) and many other factors. Nevertheless, they are most readily distinguishable features due to their very dark hues.

In late winter most of the notable rivers are pretty full of water. None of the rivers are much polluted or turbid. Their spectral signatures do not differ significantly except for shallowness and bottom reflectance factors, or nearly dry condition of some smaller rivers and ponds. Some tributaries in middle Narmada are ravinous, and some too full of alluvial detritus; and Dughi river meanders too much over very broad sand flood plain. Small rivers appear to cut through or superimposed on some hills and ranges. At places the river Narmada shows up some rocky out-crops in the mid-course or on the banks, as does the Bearma river also. The famous marble Rocks outcrops on the Narmada appear as white or very light cyan.

Lakes and ponds are next most important water bodies in region. The Upper Narmada basin abounds in lakes and ponds, particularly the triangular level tract between the Narmada and Hiran rivers, called Haveli plain. By virtue of these numerous lakes and ponds of all sizes, this part has become the richest cropland in the whole Scene. Their size and frequency diminish Westward. They occur in and around Jabalpur city also. They continue to occur East and North-East amid the hills and ranges of Bhanrer and Kaimur. Here, and farther West over the Damoh Plateau, there are a number of lakes or tanks made for irrigation by throwing small barrages across headwaters of some streams coursing down small topographic recesses. A sizeable, shallow lake to the south of Sagar is notable, and a reservoir on river Dhenwa on Northern slopes of Satpura Hills is conspicuous.

### *Recent Changes*

Some notable changes have come about since this pretty old data was acquired. Most important ones amongst them are the Tawa Dam (just West of the Scene) and the main East canal in the middle Naramada Basin issuing from it. More recently, another large dam has been completed on Narmada just South of Jabalpur city, as well as the Western canal from it. Lately, a reservoir has been created on Bebas river a little South of Sagar town to augment its water supply.

A sprawling military campus has come up on an upland South of Bebas river as an annex to the existing Cantonment at Sagar. Completion of direct Bhopal-Jabalpur route (N.H. 12) passing all along the Northern flank of the Narmada basin has brought about some land use changes and other developments in this belt. Other secondary developments (both, constructional and destructional) like new colonies and roads, enrichment of cropland, some loss of forest or scrubs etc. is also noticeable on the more recent data. An industrial township (for cement production) a little North of Damoh town and scars of open-cast mining of limestone nearby is also notable feature. The Sagar lake has lately become much shrunken, shallow, and covered with water hycaena. Around Sagar residential developments on the Eastern fringe and on the South-Western fringe can be noted on the recent imagery of Sagar. Sort of rudimentary ring roads—one from East via North to the West, and another from East via South to the West can also be seen.

### *Land Use Regionalisation*

The foregoing land use interpretation seen in the light of general geographical background of the area, particularly lithology, topography and drainage, renders the entire Scene divisible into following regions and sub-regions (Fig. 1+4)[4]:

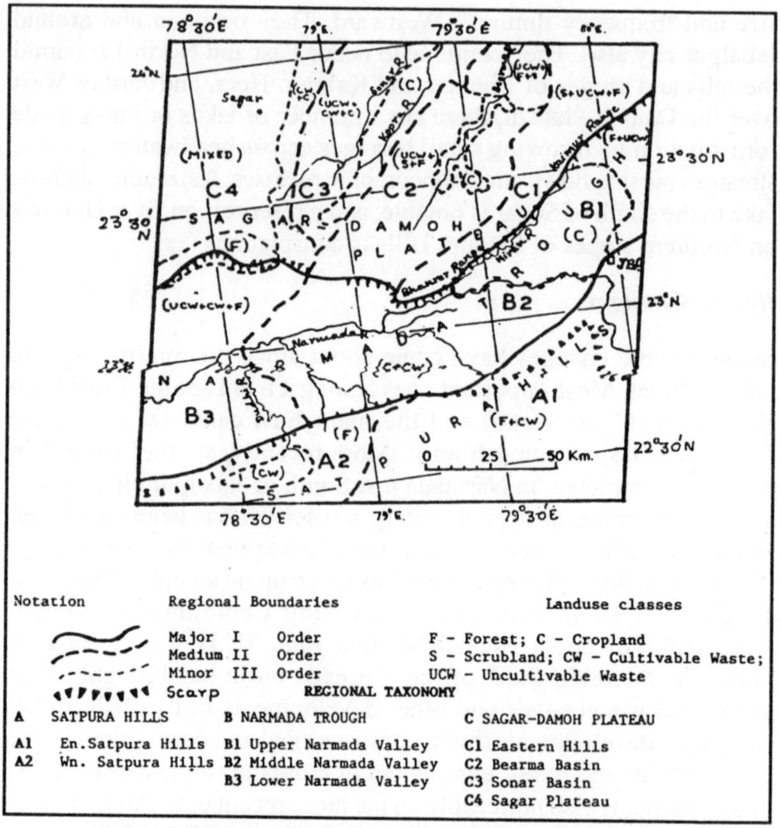


**Fig. 1 (1+4):** Geographic Framework and Landuse Regions.

**Land Use Regionalisation of Jabalpur Scene**

| *A. Satpura Hills* | *B. Narmada Basin* | *C. Bundelkhand Plateau* |
|---|---|---|
| A-1. En. Satpura Hills | B-1. Upper Narmada Basin | C-1. Eastern Hills |
| A-2. Wn. Satpura Hills | B-2. Middle Narmada Basin | C-2. Bearma Basin |
| | B-3. Wn. Narmada Basin | C-3. Sonar Basin |
| | | C-4. Sagar Plateau |

*Note*: Fig. 1+4 combines, in a single diagram, Geographical Framework of the Region (Fig. 1) and Land Use Regions (Fig. 4).

*A. Satpura Hills*

This mountainous Southern part of the Scene is largely under forests much degraded particularly in the Eastern part, and wastes. As a complex mountainous region with many river basins, it shows good

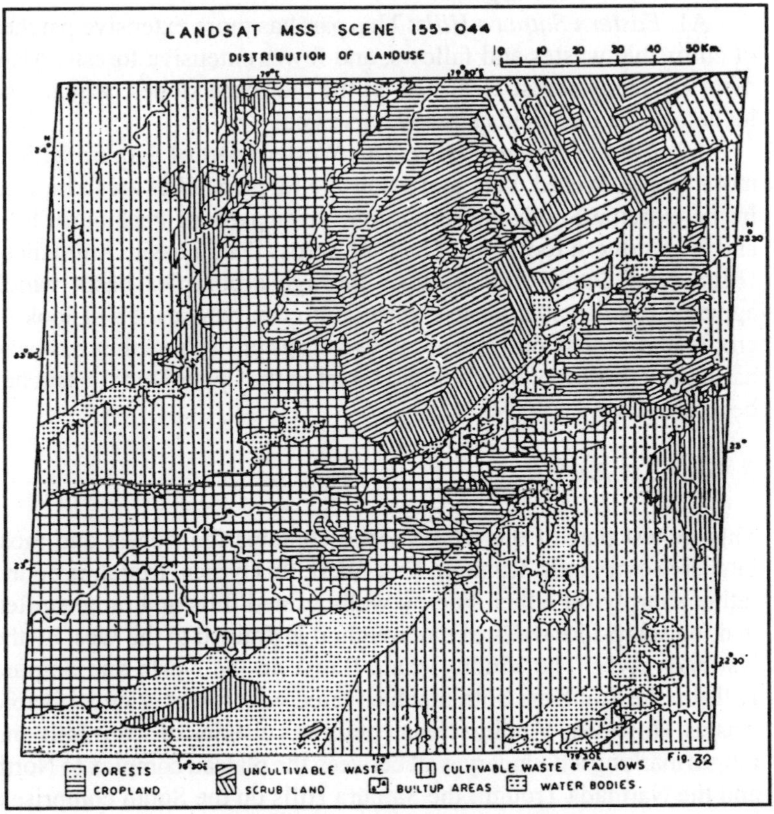


**Fig. 2:** Land MSS Scene 155 < 044

forests on the higher hills, open or degraded forest and scrubs or clearing on the lower slopes, and lean cropland in the river basins. Forests are healthier and more extensive on the Western higher hills of Satpura. Made of granites and gneisses of the Archaean group and topped with Deccan Traps on the higher Southern part, it has more forests on the Southern higher hills, whereas lower Northern slopes and basins are full of wastes, forest clearings and fallows. In this predominantly kharif crop area much of the cropland already harvested by this time may pass under fallows. Some areas show very faint crop, perhaps because Jowar and Tuar are grown together in the same fields, and by this time the kharif crop of Jowar is already harvested whereas the rabi crop of Tuar is still standing in the fields. Satpura Hills region can be sub-divided into two sub-regions:

A1. *Eastern Satpura Hills*: This part has more extensive patches of cultivable wastes and fallows, and fewer extensive forests. Also, there are many forest clearings on the Northern slopes, which also include some rice fields already harvested.

A2. *Western Satpura Hills*: This Western part of Satpura Hills is more forest covered. Very and extensive forest patches are to be seen in the extreme South-Western corner over parts of Pachmarhi hills and the Northern foot hills of Satpura across Dudhi basin. Amid these Hills upper Dudhi river courses through a complex basin which appears as cultivable wastes with faint crop patches. Faint look of cropland is mostly due one of the two crops grown together has been harvest by this time. Much of the forest in the upper Dudhi basin has been cleared.

### *B. Narmada Basin*

This richest part of the whole scene contains the richest and most intensive croplands. Running East-West, this typical trough is bound rather sharply on either side by well-defined hills and ranges often making scarp-like walls of the trough; it is drained to the West by the Narmada river. The Vindhyachala Range in the North and Satpura Hills on the South are both much covered with forest or uncultivable wastes. However, Vindhyachala comprises a series of scarp hills and ranges marking a boundary wall between the plateau complex to North and the Narmada Trough; the Satpura Hills on the South comprise a broad mountain complex whose Northern slopes are comparatively gradual and much eroded too. Hence, these broader slopes are more degraded forests, scrub or cultivable wastes. The Narmada river itself flows comparatively closer to the Northern wall of the trough, leaving a comparatively broader Southern flank of the trough obliquely traversed by numerous tributaries for the Satpura Hills. Trunk routes of the trough as well as a larger number of towns and villages are located in the Southern flank of the basin. It is divisible in three sub-regions:

B1. *Upper Narmada Basin*: After traversing a long circuitous course the upper Narmada river enters the structural Narmada Trough/ Basin from South-East and courses West-North-West to reach very close to Northern wall of the trough and takes an essentially Westward course. Its most important right-hand tributary Hiran river also first courses west well nigh to foot of the Bhanrer Range and turns sharply South-West to follow a long way very closely along the long Bhanrer range to pour into the Narmada where the latter comes closest to the Northern wall. The two rivers thus define and enrich an extensive triangular level plain between them. It is beset with numerous lakes and ponds all over which augments the rich alluvial soils of the area. This makes it the richest and most extensive cropland of the whole Scene. Supported by extensive irrigation and/or soil moisture, the rabi crop (mostly wheat) appears very luxuriant in this season.

B2. *Middle Narmada Basin*: Extending over the middle stretch of Narmada basin, this region shows typical land use of rich river plain. Flowing mostly pretty close to the Northern wall of the basin, the principal river is joined by several notable streams from south, which bring copious alluvial detritus and ground water also. The areas closer to main and the tributary streams are rather kharif cropland (mostly sugarcane) which been harvested, and the stripped fields look compounded with cultivable wastes and fallows. Upland and drier parts a little away from the stream channels are under rabi crops, and look luxuriant in this season. Some crop areas look very lean due to one of two crops grown together having been harvest by this time.

B3. *Western Narmada Basin*: West of about 79° East longitude, the geomorphic and land use character of Narmada basin changes significantly. The Northern wall of the basin recedes back substantially from about 78° East to just past 79°East longitudes and there are notable rivers coming from North. All these Northern tributaries and the Narmada itself are notably ravined and deep-cut. Over about the mid-run of this stretch of the receded Northern wall and a fairly broad Northern flank of the basin, however, there arises another sharp, narrow range or ridge between the trunk stream and the Northern wall across these courses which appear to be cut-through or superimposed on this Udaipura ridge. While the Ridge itself is covered with degraded open forest or uncultivable wastes, there are good patched of cropland between the deep-cut valleys, both North and South of the Udaipura ridge. In the lower courses, these tributaries run sub-parallel and close to the trunk stream. Much ravined as they are, they appear as conspicuous belts of cultivable wastes, while away from stream channels there are good crop patches with a purple overtone due to black soils character.

There are many more streams coursing down from South, bringing down huge amounts of alluvial detritus and spreading the same as broad white belts of sand along their meandering lower courses. Lower Dudhi is truly conspicuous by extent of meandering over the sandy belt of its channel. White sand patches often with intricate channel patterns cover a large proportion of areas as sandy wastes. Tough Narmada river is also much ravined hare, some of the wastes and fallows are harvested kharif crop areas. Crop area here also show the purple overtone of black soils.

*C. Bundelkhand Plateau*

Made of middle Vindhyans—mostly softer Shales in the river valleys and much harder Sandstones/Quartzites as hill-ranges. This plateau complex is gently tilted Northwards, and has its grain and drain in the same direction. Western part is overlain with basaltic Deccan Trap which thins out to the East. The whole plateau complex is comparatively a dry area. While in the Deccan Trap, the Shales in the river valleys particularly support good vegetation growth. Sandstone and quartzite uplands and hills are comparatively dry rocky areas with less vegetation cover. This huge area is divisible into following four sub-regions:

C1. *Eastern Hill-Ranges*: This North-Eastern-most part of the Scene is complex area with hills and ranges trending North-East and roughly parallel to the Bhanrer and Kaimur ranges, some of which provide holding wall to many a sizeable lakes and tanks on their one side, such as Borhribandh reservoir. This very complex hill area has distinct cores of forest covered hill-ridge summits surrounded by belts of rocky uncultivable belts, and then by scrubs on the periphery. Or sometimes directly scrubs belts. Other land uses are minor, and it contains a sizeable proportion of scrubland. The extreme North-Eastern parts show a light purple over-tone due to meta-sedimentary character of the rock formation here.

C2. *Bearma Basin*: Also called Damoh Plateau, the whole upper Bearma basin in the scene is a rugged, rocky, and largely barren country; and uncultivable, rocky wastes predominate the land use pattern. Uppermost reaches and peripheries of the Bearma basin dramatically reveal the numerous fine fractures criss-crossing the country there as fine red lines, because fractures do hold some soil moisture and support some vegetation. At the junctions of such fractures, there are specks of red and small village nearby each. The Bearma river itself supports some rabi cropland patches in its closer vicinity, surrounded by a broad zone of uncultivable rocky wastes, particularly, on the eastern flank.

C3. *Sonar-Kopra Valley*: The valley of Sonar river with its notable right bank tributary Kopra river makes the richest sub-region North of the Vindhyachala Range with extensive cropland particularly in the inter fluvial tract between the two rivers. East of the principal river Sonar, there appear some rocky outcrops. An equally sizeable left bank tributary of Sonar river, e.g. Bebas river, however, does not show as rich a cropland as does Kopra, which courses largely through a less congenial rock formation, the Lametas. In the upper reaches of the Sonar valley, the cropland fades, giving place to cultivable and uncultivable wastes and minor patches of forests also in the higher most areas. While on east margin the region merges rather gradually with the Bearma Valley, the Western margin is marked rather more abruptly by a belt of cultivable and, sometimes, uncultivable wastes.

C4. *Sagar Plateau*: This North-Western-most part of the Scene shows a juxtaposition of Trap cover and exhumed Sandstone outcrops, and has much undulating topography. Though it shows rabi crop patches far and wide, the land use pattern here is actually intricately mixed and complex, where all other land uses are intermingled with crop patches. Sagar town with an extensive shallow lake to its South-east looks quite distinct. Small rivers like Bebas on the South-eastern margin and Dhasan on the North-West do not support extensive cropland. While the North-Western corner has much open forest and uncultivable wastes, the Eastern and South-Eastern margin of Sagar Plateau is marked clearly by narrow belt mostly of cultivable wastes, which coincides with the Lameta rock formation here.

## *Geologic-Geomorphic Controls on Land Use*

The foregoing interpretation and analysis of land use and settlements substantiates the hypothesis that the geologic-geomorphic factors of the area bear profound influence on the pattern of land use and settlement, more so in the areas of the varied geology and topography.

Significantly controlled by such factors, the land use pattern in this area can be represented as a cross-section to best reveal this control—a hypothetical cross-section taken in the North North West—South South East direction, i.e. across the general grain of the country[5]. A secondary profile is similarly drawn across the Bundelkhand Plateaux in a West North West—East South East direction. In this respect, following points are pertinent (Figs. 3A and 3B)[6].

Higher elevations, like summits of hills and ranges, are covered with forests; their lower slopes, where forest have been cleared or degraded, appear as scrub wasteland. Very seep brows appear as uncultivable rocky wastes. Satpura range and the Vindhyachala hill-range in North-eastern corner of the Scene reveal these points.

In the structural trough, the Narmada mostly flows closer to the Northern wall of the trough and receives much fewer tributaries from that side. Its numerous tributaries flowing from Satpura Hills from the South bring great amounts of sandy sediments in the trough. Thus, areas close to the numerous river channels appear as sandy belts of cultivable wastes and fallows. Beyond such channel belts, the copious supply of fresh alluvia and groundwater helps this area support rich croplands. Such croplands appear more broken in Western part where river are ravinous also. In the Eastern part of Narmada Valley agricultural richness is further augmented with occurrence of numerous ponds.

*Extensive Northern part of the Scene*—North of the Vindhyachala Range—comprising the Bundelkhand Plateau complex is essentially a Vindhyan country overlain in the Western part with basaltic Deccan Traps. The Vindhyan formation broadly consists of softer Shales facilitating development of broad river valleys, and the much harder Sandstones/Quartzites forming interfluvial hills and ranges. As the whole formation is gently tilted North wards, and the structural and topographic lineaments mostly trend to North or North-East, this whole country is much undulating from east to West. The land use likewise varies in the same direction (Fig. 3). Black soil areas have a darkish and bluish-purple overtone added to other tonal hues of those areas.

Basaltic Trap cover has contributed black loamy soils of varying depths in the plains and valleys and thick, bouldery hill tops, which are more congenial to vegetal growth compared to the typically Vindhyan country. This would manifest in the richer forest and croplands in the Western basaltic parts. However, this factor is countered

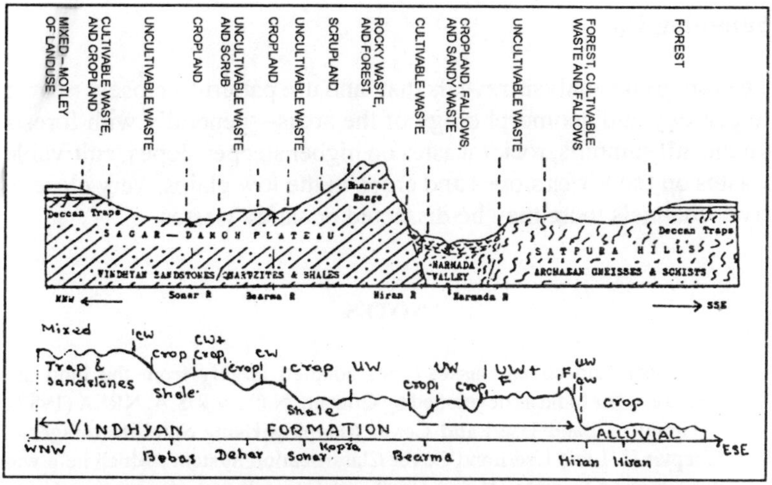


**Fig. 3:** Schematic Geologic-Geomorphic and Land use Profiles: (A) in the NNW-SSE direction, and (B) in the WNW-ESE direction. F-Forest; CW-Cultivable Waste; UW-Uncultivable Wastes.

to some extent by an increase in the total rainfall to the East. Only, the multifarious lithologic composition with basaltic hills/black soils, exhumed sandstone hills and scarps, have contributed to an intricately intermingled land uses, particularly in the North-Western part. Higher parts in the inaccessible areas have open forests, scrubs or stony wastes, and lower parts have cultivable wastes or cropland. About a score of km. East of Sagar, there is long narrow belt of Lameta formation running North-South, which is largely covered with cultivable wastes.

If Sonar-Kopra have richer cropland, Bearma basin in the Eastern part of the plateau is a more rugged and stony country with extensive uncultivable wastes. In the stony upper Bearma basin, there are numerous small fractures cris-crossing the country. In the dry, stony background these fine fracture lines holding some groundwater and supporting some plant growth are dramatically revealed in the satellite imagery. Junctions of these fracture lines are hosts to croplands and settlements. Very close to the river itself, Bearma valley has patches of cropland surrounded by stony wastes farther a field.

In the extreme North-east corner also the lithologic character changes to meta-sedimentries of Bijawar group, which shows a light purplish over-tone added to various hues.

## Summing Up

The foregoing analysis reveals that land use pattern is closely related to geology and geomorphology of the areas—generally with forests on the hill summits, rocky wastes on higher steeper slopes, cultivable wastes on the lower slopes and crops in the low plains. Very close to river channels there may be dissected or sandy wastes.

### NOTES

1. Land use land cover classes/types adapted largely from the land use taxonomy for India as developed by Goutam, N.C., N.R.S.A. NRSA (1987), 'Manual of Land Use/Land Cover Mapping Using Satellite Imagery', Chapter II 'Land Use/Land Cover Classification System'. Much help was also derived from his 'Land Use/Land Cover Interpretation Key Using Satellite Remote Sensing Imagery'. In India pasture and grasslands is not a very significant or extensively occurring land use/land cover category. As such, it is treated merged with cultivable wastes in this study.
2. Fig. 2 shows many areas with two land use types dominating some areas, by super-imposing two types of hatching. However, in such cases, the line or dot spacing of superimposed hatching has been doubled in order to balance the density of hatching of various land use regions. As such, there are more types hatching patters on body of the map than are shown its legend.
3. At the height of winters the difference between the Wet and the Dry deciduous types of forests is not visually apparent at this low resolution, small scale data.
4. Land use Regions (Fig. 4) as shown combined with Geographical Framework of the Region (Fig. 1), specifies the dominant land use types of each region with abbreviated name symbols, which are written in the order of dominance of land use in each region.
5. Help and suggestions from Geologists, Prof. A.S. Dave, Pt. R.S. Shukla University, Raipur, and Dr. P. Kathal, Dr. H.S. Gour University, Sagar in this respect are gratefully acknowledged.
6. Fig. 3 presents hypothetical geologic-geomorphic and land use profiles/cross-sections approximately in the A. the NNW-SSE direction across all major units; and B. the WNW-ESE direction across the Bundelkhand Plateau which shows main variations in that direction. They purport to present a model of Geologic-Geomorphic factors and their bearing on land use and human settlement.

# 5

# Satellite Remote Sensing and GIS for District Level Planning in India
## *A Case Study*

P. S. RATHORE

The development of any country depends on the effective and optimum utilization of its various natural or physical resources. As early as in the 1960s, India realized the importance of remote sensing technology and its role in natural resources mapping and monitoring. Consequently the Indian Remote Sensing program was initiated in 1970. Modern remote sensing technology with its new detectors and imaging technologies provides a number of efficient and consistent ways to gather timely spatial data of the entire globe, at scales appropriate to a variety of applications.

Satellite remote sensing technology plays a vital role in formulating a developmental plan at national, state and micro (district/ block/taluk/mandal/village/watershed) level. Using this new technology not only does the planner acquire the required information/ data fast, but is unbiased and cost effective as well. Thus, the accuracy of plan can be improved very significantly. Remote sensing technology can give information on most of the natural resources like soil cover, vegetation cover conditions, geological and geomorphological information, groundwater potentials, flood risk management and drought prone region studies, land use/land cover etc.

India is now at a fairly advanced stage and in good command of receiving data from several indigenous and foreign satellites, equipped with wide range of sensors and data acquisition aids. Remote Sensing data collection is valuable aid to know our valuable resources and provides us data that can be used for various developmental activities and planning of existing resources in conjunction with the ground truth surveys. It helps us to survey various natural resources like land use, agriculture, forestry, minerals, water etc. and study various physical phenomena. Remote sensing is a multi-disciplinary technological system. It relies heavily upon physical means mathematical analysis and bio-geosciences rules which enables us to understand various conditions, characteristics and changes of the atmosphere, hydrosphere, biosphere and lithosphere.

GIS is a system for storing, retrieving and analysing geographically referenced data sets. In GIS, spatial (e.g. maps, satellite imagery) and non-spatial (e.g. census data, field data) sources can be integrated and a set of spatially registered data layers can be analysed independently or in combination with a number of layers. Thus, one can stitch together bundles of data from a wide variety of sources and manipulate them to get quantitative information for resources management and planning at different levels. GIS consists of such activities as capturing of data, archival, new survey data, integration of data sets, design of efficient data structure, database, user interfaces and analysing integrated data sets. The ability of GIS to identify new relationships from the existing database elements and to present the results of the analysis in cartographic form makes them exceedingly important from planning point of view, as most macro and micro-level planning tasks ideally require geographically referenced information.

In India for long, planning and decision-making functions remain very much centralized and vertical around two nodes viz., the union and the states. Urban and rural development bodies mostly act as civic functioning bodies and not as instruments of micro-planning. District planning occupies the third ring in the planning hierarchy below the national and state levels. Its position has become more strategic, as it is at the district level; we possess the basement of the planning mechanism. The block and village levels plans are of vital importance, the country does not yet possess the requisite machinery for planning at such micro-levels. Great significance is attached to district level planning as the districts are to plan for their development considering their own Agro-climatic and socio-economic conditions and while doing so, they should reflect the basic objectives of national plan, viz., (i) Conservation of resources, (ii) Increasing production, (iii) Reducing unemployment, (iv) Alleviating poverty and (v) Environmental issues.

**Importance in Indian Scenario**

In our country the greatest limitation to well guided district planning exercise has been the absence of an integrated database of physical features and natural resources either at macro or micro level. The conventional methods of obtaining essential data on physical, geographical, natural resources and land use data and maps were extremely slow wherein; the rapid changes could not be brought out. In India the importance of district planning arose primarily out of the following pragmatic considerations, viz.

- Block Development Programme (BDP), which failed to make an impact on grassroots development, though block development officers knew the problem of the people.
- Officials at grassroots level lacked the planning perspective, authority (both administrative and financial) and also the technical competence to formulate plans and implement them on their own.
- Subsequently, Community Development Programme (CDP), which established a coordinated staffing pattern, did not have a plan for integrated resource development.
- Macro plans for Planning Commission also failed to make the requisite impact on the grassroots level because they are too far removed from the local resources and realities.

Thus, the idea of district planning has emerged as an attempt for development from below with an emphasis on the removal of disparities. Conceptually, districts are more homogeneous unit areas than states from planning point of view.

Planning at district level plays a vital role in the economic growth and the development of the area, more so, the degree of development is a function of natural resources occurrence, its utilization and sustenance for future population needs and demands. Today to meet the growing demands of teaming millions these resources are being restored to intensive utilization resulting in decrease in vegetative cover, loss of forest wealth, transformation of production agricultural land into degraded land dwindling of water resources, environmental pollution due to rapid urbanization and industrialization diminishing per capita availability of land, famine and drought, migration of labour etc. Up-to-date and reliable information about the use of various natural resources and socio-economic features is essential for the proper management, regulation and planning of such resources. If the country is to overcome the problems associated with haphazard and uncontrolled growth there is an urgent need for a sustained effort to reverse the situation to normalcy through optimal district level planning and scientific management of all the natural or physical endowments.

**Review of the Work on the Subject**

The history of planning can be viewed in terms of distinction between certainty and uncertainty, simplicity and complexity (Michael Batty, 1979). Planning is defined as the process of 'linking knowledge to action' (Friedman, 1976). The planning process based on the systems approach represents a conveyance of ideas from several sources from system theory itself from scientific method, from design method and from the various decision-making problem solving methods adopted (Brain Gordall and Andrew Kirby, 1979).

According to Planning Commission of India, planning involves the acceptance of a clearly defined system of objectives in terms of which to frame overall policies. It also involves the formation of a strategy for promoting the realization of ends defined. Planning is essentially an attempt at working out a rational solution of problems, an attempt to coordinate means and ends, it is thus, different from the traditional hit-and-miss methods by which reforms and reconstruction are often undertaken (Planning Commission Report, 1967).

In our country we have multi-level planning i.e., the plans are prepared at the national, state and district levels. The process of planning essentially means the applications in a comprehensive manner of scientific methods in dealing with problems of economic development and keeping in view the inter-relations of different sectors as well as the needs of future (Venugopal Reddy, 1979). Realizing the significance of planning, subsequently the idea of decentralized planning arose. Need for decentralisation was felt with one reason bearing importance like "planning has to take into consideration the existing ecology since our economy is basically agricultural". The agro-climatic zones within which the district needs to be identified, characteristics and needs properly reflected which would help in agro-sectoral plan of the district, which can further be detailed into the state plan (Mary Parmar, 1992).

District level planning a concept of decentralised micro-planning, is an area based sub-state planning aimed at supplementing the national and state plans with a more detailed examination of resources, constraints and potentials of local area. Drawing up a district plan requires up-to-date and reliable information on natural resources in spatial format at desired scale. Thus, arriving at an alternate planning scenario requires integration of various sources, which may be in different scales, format and projections. In addition, it is also required to relate the spatial and tabular data for proper planning.

Noteworthy efforts in field of planning were made by D.R. Gadgil, an eminent economist and planner. Balwant Rai Mehta's 'democratic decentralization' gave much impetus to the concept of decentralized planning. Ashok Mehta Committee (1977) too wanted the decentralization in power and planning at district level, it was given due weightage by this committee. In 1969, Planning Commission had framed guidelines for district plans. The working group of Planning Commission in 1977 under the Chairmanship of Prof. M.L. Dantwala drew up guidelines for planning at block level. In September 1982 working group on district planning was set-up by the Planning Commission.

As geographical factors greatly influence the development and growth of any particular area, thus, planning at district level adds a great deal in way overall development of the area. To this much of the research work was taken up using conventional and statistical methods are still continuing along with modern techniques. Different organizations, institutions and individuals at different capacities are working to bring out studies on land use and planning at different levels.

Reliable information and data on resources is essential for effective management of the natural resources of the country. The role of remote sensing data in providing such information has been well recognized, Ali (1949) described about the land utilization survey in various countries like U.S.A., Great Britain and China. He suggested the method followed in U.S.A. to be more suitable for India.

With the advent of remote sensing techniques in 1960, land utilization particulars were gained using traditional ground methods using aerial photographs by the Survey of India in 1960, particularly for urban studies the new trend started. After the launch of LANDSAT-1, the satellite remote sensing techniques from 1972 were increasingly being used for survey of different natural resources.

Indian Institute of Remote Sensing (IIRS), Dehradun of National Remote Sensing Agency (NRSA) carried out an inventory of natural resources of Karimnagar district and prepared land use, land resources maps along with other thematic maps. At this time some of the research work conducted by Gautam, N. C. (1982), Dhabariya, S. S. (1988), Rathore, N. S. (1988), Muralikrishna, I. V. (1992), Singh, S. (1992), Sharma, J. R. (1994), Behara, G. (1994), Manchanda, M. L. (1995), Hegde, V. S. (1995), Perumal, A. (1995), Raghavswamy, V. (1981), Nagaraja, R. (1989), Ravi Shankar, G. (1992), Saxena, M. R. (1993), Suresh, L. S. (1997), Jayanthi, S. C. (1997) and Bhudevi A. (1998).

Different institutes at national level and regional level such as National Remote Sensing Agency (NRSA), Hyderabad (AP), Space Application Center (SAC), Ahmedabad, Regional Remote Sensing Service Centers, B.M. Birla Institute of Science and Technology, Jaipur (Rajasthan) are some of them engaged in research and planning at different levels using the advanced technologies of Remote Sensing and Geographic Information System (GIS).

National Remote Sensing Agency (NRSA), Hyderabad has taken up number of remote sensing projects on behalf of many Central and State Government departments and other private organizations/ institutions in the country and prepared various thematic maps on various scales in field of geology, soils, hydrology, wastelands and land use etc. Particularly in the field of land use studies and wasteland demarcation NRSA has carried pioneer works.

Apart from NRSA projects, many studies on regional, district, panchyat and mandal planning and sustainable development employing remote sensing techniques were undertaken by various scientists in different parts of India and abroad. Variety of published work of researchers, scholars and scientists in various books and journals also focus the amount of research work conducted in regard using techniques of Remote Sensing and Geographic Information System (GIS).

In spite of all the above-mentioned attempts, there still lies vast potential for research work to be conducted at micro-level planning for overall development of any area.

**Objectives**

The objective of the present work is to design and prepare a developmental plan for District Rajsamand of Rajasthan with the view to support the decision-makers at the district level involved in the district planning exercise. A computerized information system prepared with the extensive use of Satellite Remote Sensing and Geographic Information System (GIS) techniques would be aimed at the broad objectives mentioned below:

1. Development of geographical information system for district level planning integrating the information on natural resources and socio-economic features.
2. To develop a systematic procedure of data integration for development around a GIS package.
3. Assess the utility of Satellite Remote Sensing data and prepare different thematic maps on 1:50,000 and 1:250,000 scale.
4. Identification of resources management strategies to meet the basic needs of the people from backward group.
5. Identify and suggest measures to increase productivity for the economic growth of the district.
6. Recommending a developmental plan keeping in view the intricate ecological balance.

## Methodology

A detailed inventory of natural resources along with the socio-economic conditions is a prerequisite to achieve target-oriented goals in land development in general and planning at district level in particular. To this an attempt is made to follow a standardized methodological framework well approved in the scientific world. The procedural steps involved in the study are presented in Fig. 1.

## Data Integration Using GIS

The process of unifying existing data sources into a single framework is called database integration; it is one of the most sophisticated and powerful approaches to data interpretability, where a complexity made manageable. "One of the major problems that the development of Geographical Information System (GIS) applications are facing today is data acquisition. Not that data is not available: geographical data collection has been going on for centuries. Some of that is still stored on paper, including maps, some has been digitized and some is stored

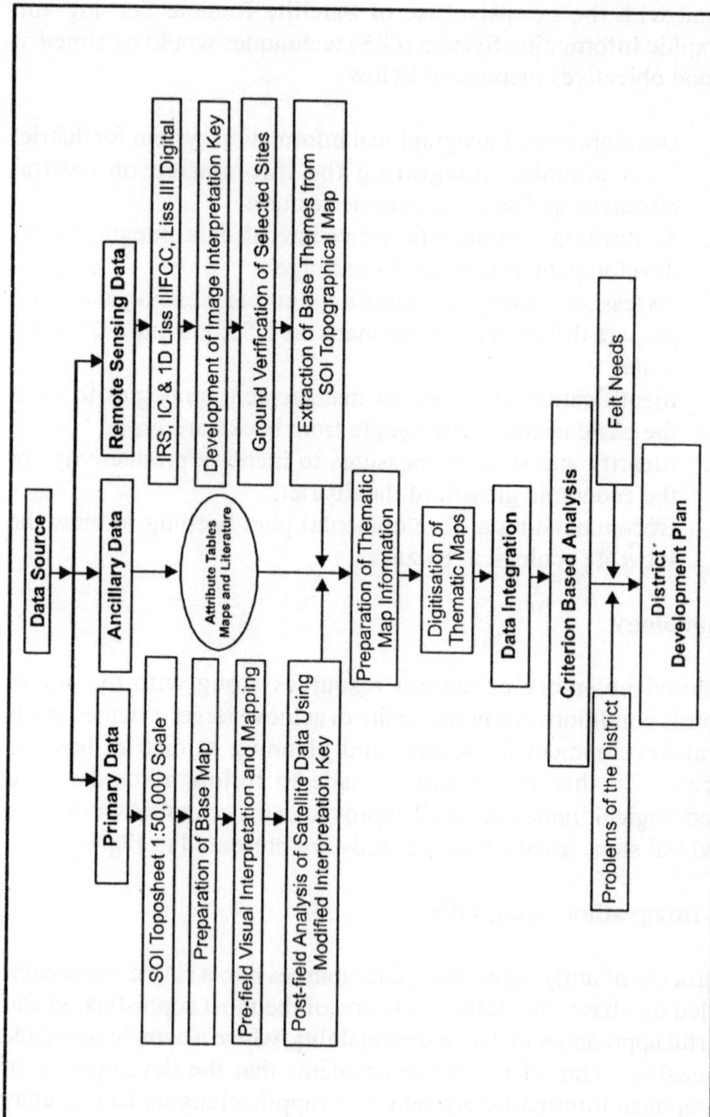


**Fig. 1:** Procedural Steps to Generate District Development Plan.

in current GIS (at best) or in traditional files or databases. However, too often their reuse for new applications is a nightmare, due to poor documentation, obscure semantics of data, diversity of data sets (what information is stored, how it is represented and structured, what quality it has, which date it refers to, which scale is used...), and the heterogeneity of existing systems in terms of data modelling concepts, data encoding techniques, storage structures, access functionalities, etc." (Thomas Devogele *et al.*, 1998).

Spatial information in the form of resources maps viz., Land Use/ Land Cover, Soil, Hydrogeomorphology and Ground Water Prospect, were prepared using satellite remote sensing data using Visual Interpretation Techniques. Other thematic maps viz., Settlement Location and Transport Network, Administrative Boundaries, Drainage and Surface Water Bodies were prepared using the base details from Survey of India Topographical Maps on 1:50,000 and 1:250,000 scale. A Slope map was prepared using the Contours and Spot Height of SOI topographical maps, utilising the capabilities of ArcInfo software.

## Data Base Design, Creation and Organization

In order to meet above-mentioned objectives, the whole exercise calls for a planned and rationally structured database. To this it becomes imperative to segregate the database into two parts, namely, Spatial Database and Non-Spatial Database. The spatial database contains a store of all map information derived from various sources, whilst the non-spatial database basically stores alphanumeric and all other forms of data, which did not get covered in the above. As spatial analysis forms the backbone of this whole exercise, the data storage facility also required to be up-to-date and efficient for quick data manipulation.

### *Database Design*

The very first task in the process of data base design is the identification of complete set of primary input data elements. This was done keeping in view the non-redundancy in the data input as well as multiple usage of single data element for different purposes. Input data elements are available in two forms viz., spatial and non-spatial. Design of the database for district planning had to recognize the fact that the system has to support information retrieval in terms of spatial units, which are generally used by planners at district level. The spatial inputs are generally referred to SOI topographic Maps. An administrative unit may include partially or fully, more than one map sheet. Tabular data is referred to either point locations or village/ taluk names. Keeping in view the database has been designed in such a way that it facilitates basically physical (schema) set, which is adequately supported by logical set up.

*Physical Schema*

It is the system view providing the basic framework for database creation and management. In this definition the actual data elements are defined uniquely. Physical schema consists of graphic and tabular components. Graphic components include a set of thematic layers used for different purposes. Each layer has a set of common TICS/ registration marks, which are the inter-section points of latitude and longitude lattices. This is essential for bringing the different thematic layers onto same coordinate database creation from multiple map sheets by providing the flexibility of individual map inputs. Tabular data elements include the description tables for the graphic features stored as a part of graphic layers as well as data on socio-economic aspects.

***Database Creation and Organisation***

In order to have the database ready to use, the main tasks carried out are:

- Collection and/or preparation of input data sets.
- Database creation and its organisation into logical and physical schema as detailed earlier in this chapter.
- Transformation/Analysis of primary input data in order to generate desired information and organisation of information into different set up.

This exercise involved collection of the data already existing with the agencies as well as certain data identified to be generated from interpretation of remote sensing data. Remote sensing inputs included land use/land cover map interpreted from IRS-1C, LISS-III data of

October1998 and February 1999 at 1:50,000 scale. A hydro-geomorphology map, prepared using remote sensing data was used. This was a map prepared after visually interpreting IRS-1B, LISS-II data of October 1988 and January and February 1989 at 1:50,000 scale with limited field checks. Apart from this a soil map was also used on 1:250,000 created by NBSS, Nagpur.

Inputs from SOI toposheets on 1:50,000 and 1:250,000 scale included elevation counters and Spot heights for the preparation of Slope Map, Drainage and Surface Water Body Map, Settlement Location and Transportation Network Map and Administrative Boundaries (District/taluk).

Inputs collected from state departments include maps/tables of villagewise amenities availability and climatic data. Census handbook was used for socio-economic data.

Thus, the data sets collected/prepared through multiple sources and methods were put to GIS environment. To this, in the present study, Arc/Info (ver. 7.2) GIS package has been used as the core of the spatial database. It is a modular and vector based package where the spatial data is organized using topological data model while the non-spatial data is stored using a DBMS package (the capabilities of the package are described in previous chapter). The data entry was via scanning and digitization (in case of Map inputs using the capabilities of Arc/Info and AutoCAD MAP) and directs entry through alphanumeric terminal in case of tabular inputs.

*Creation of Spatial Framework*

The most significant task prior to map scanning or digitization due to the fact that entire district is covered in multiple map sheets and inputs were on single map sheet basis. It involved:

- Definition of a coordinate system with origin lying at lower left corner of the lower left map sheet.
- Identification of registration tic marks at latitude-longitude crossings.
- Design on coordinate units chosen as true distance in kms. on ground.
- Calculation of coordinate values for selected registration points using polyconic projection with appropriate central meridian and standard parallel.

The digitizing errors have been corrected using ARCEDIT module. Six registration points (TICS) have been used for registering the digitized map to the ground, and to transform the coverage from digitizer units to real-world coordinates.

There are 989 inhabited villages in district Rajsamand as per 1991 census. A village boundary map at the scale of 1:50,000 was prepared for all the blocks of the district, to this the census maps has been scanned and digitized using AutoCAD MAP and Arc/Info (ver. 7.2.1) GIS software. The non-spatial data used for these villages include demographic and socio-economic data, it is discussed in detail in the subsequent chapter.

An important aspect of integrated area-level planning is the combined analysis of the thematic natural resource data and the tabular socio-economic data. For this, a criteria based analysis of the integrated databases has been carried out in GIS environment. User interactive program has been developed in GIS using Arc Macro Language (AML) of Arc/Info GIS Software.

Detailed surveys, until 1960s, have been mainly carried out in developed countries and resource maps have been produced for a range of purpose with a wide variety of scales and classifications. The primary purpose of these surveys has been to determine the spatial distribution of various categories at a particular time. The resultant maps have provided planners with useful analytical tools for general or reconnaissance evaluations as well as establishing a permanent database as a part of a continual monitoring of landscape. These surveys have been carried out using time-consuming and often expensive methods involving the collection of information from various agencies, field reports and other ancillary data.

It is here that Remote Sensing and Geographic Information System played a vital role in generation of various thematic maps. The essentiality of resource maps for decision-making became must in the present world. Utility of resources maps can be summarized as below:

- Resource maps provide a basic data on type and spatial extent of resources at National/State/District/Village level for management and planning.
- These resource maps can be used in integrated manner for making development plans.
- The information on resources will be useful in assisting various programmes such as reclamation of degraded lands, floods,

ground water, mineral targeting, forestation programmes, drought studies etc.
- Resource maps can be used for subsequent updation of details and their monitoring periodically.
- For Geographical Information Systems development.

**Socio-economic Data Input and Analysis**

In order to demonstrate the over all socio-economic status of district Rajsamand different Demographic and Economic indicators were shown in Map form which would further help for better understanding of the current phenomena in the district. It is described in detail in the subsequent paragraphs with different maps.

***Derivation of Spatial Patterns—Development Levels***

Demonstration of the utility of a GIS to represent the non-spatial data spatially, different non-spatial parameters has been used to derive spatial patterns of village development levels. These spatial patterns highlight the low development areas in the district and also potential in showing the trend of development which is some of the pertinent spatial patterns are generated in this study (Fig 2). These spatial patterns have been represented on a village-wise basis and to this a village boundary map was prepared at the scale of 1:50,000 scale and Census data at village level were linked using the capabilities of the INFO Module of Arc/Info Software.

**Integration of Spatial and Non-spatial Data—*Generation of Land Resources Development Plan***

Through systematic analysis of land use/land cover, slope, hydro-geomorphology and soil coupled with agro-climatic and socio-economic profiles suitable combination of land use practices have been recommended for land resource management on a long-term basis. The basis of these development packages can be grouped based on different geophysical and environmental factors, viz. landform, slope, soils, rainfall and other climatic factors and human induced factors like land use, irrigation practices, infrastructure development and other socio-economic factors. Keeping in view the current scenario of the district as mentioned previously, the developmental

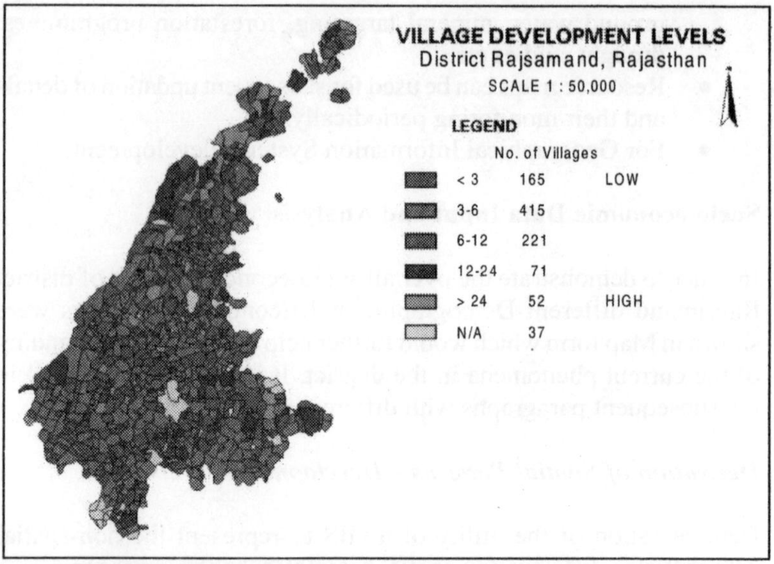


*Source*: NIC Data & GIS Analysis.

**Fig. 2:** Village Development Levels

plan for land resource management was developed, (Fig. 3) wherein following land use practices are suggested:

1. Intensive agriculture
2. Double crop
3. Agro-Horticulture
4. Horticulture with soil conservation
5. Agro-forestry
6. Conservation/forest protection
7. Afforestation with soil conservation
8. Silvipasture
9. Plantation and fuelwood trees/fast growing trees
10. Pisiculture with desiltation of tanks
11. Selective grazing
12. Protective mining
13. No action

**Results and Conclusions**

It is generally felt that one of the reasons of our failure to achieve the

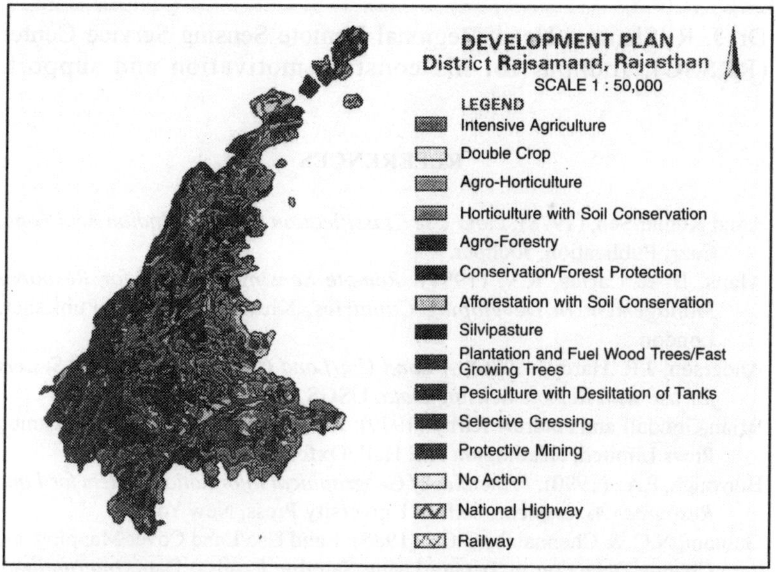


*Source*: IRS-1C (LISS-III)-1998-99, SOI Maps, Natural Resources and Socio-Economic Data.

**Fig. 3:** Development Plan

basic goals is that the plans are formulated at higher levels, which are quite distant from the grassroots realities. It is perceived that lower the level or unit of planning smaller will be the gap between planning and implementation. This is one of the strongest justifications of decentralised planning. The study has amicably demonstrates the applied concepts of organizing a systematic database of spatial and non-spatial data around a GIS core for addressing various planning issues in the different areas in a district. The experience gained from this study will help organize systematic GIS based Information Systems to combat with different problems at micro and meso level.

**Acknowledgements**

The author is greatly indebted to Dr. N.S. Rathore, Associate Professor, Department of Geography, M.L. Sukhadia University, Udaipur for his valuable guidance. Thanks are also due to Dr. N.C. Gautam, Director, Center for Land Use Management, Hyderabad and

Dr. J. R. Sharma, Head, Regional Remote Sensing Service Center (RRSSC), Jodhpur for his constant motivation and support.

**REFERENCES**

Amal Kumar Sen, (1978): *Land Use Classification System in Indian Arid Zone*, Cazri Publication, Jodhpur.

Alans, B. & Carlos, R.V. (1991): *Remote Sensing and GIS for Resource Management in Developing Countries*, Kluwer Academic Publisher, London.

Anderson, J.R. Hardy, (1976): *A Land Use/Land Cover Classification System for use with Remote Sensing Data*, USGS professional paper, 964.

Brian Goodall and Andrew Kirby (1979): *Resources and Planning,* Pergamon Press Limited, Headington Hill Hall, Oxford, England.

Burrough, P.A., (1990): *Principles of Geographical Information System for Land Resources Assessment*, Oxford University Press, New York.

Gautam, N.C. & Chennaiah, G. Ch. (1985): Land Use/Land Cover Mapping and Change Detection in Tripura Using Satellite Landsat Data, *International Journal of Remote Sensing*, Vol. 6, Nos. 3-4.

Government of India, Planning Commission, *First Five Year Plan*, p. 7.

Mahesh Chand & Puri, V.K. (1983): *Regional Planning in India*, Allied Publishers Pvt. Ltd., New Delhi.

NRSA (1991): *Technical Guidelines, Integrated Mission for Sustainable Development*.

Rathore, N.S. (1992): *Natural Resources Base Development*, Scientific Publishers Jodhpur.

Sudipto Mundle (1977): *District Planning in India,* Indian Institute of Public Administration (IIPA), New Delhi.

# PART II

# Agricultural Land Use and Development

# 6

# Agricultural Land Use, Productivity and Food Self-sufficiency in Bangladesh

MRITYUNJAY M. JHA

## Introduction

The definition of food security[1], as given by the FAO covers "physical, social and economic access to food", clearly indicates the necessity of food availability as critical factor in ensuring food security. This becomes further pronounced in agrarian economies with large population base, where agriculture is considered not only as food production system, but also as a livelihood provider to millions at broadly three operational levels i.e. production, processing and marketing. Agriculture, as a dominant sector in South Asia, touches food security at all levels, viz. individual, household, regional and national. More than 500 million South Asians live in absolute poverty. The crisis of food insecurity in the region is mostly related to low access than low availability, there is no denying the fact that the agricultural land use holds the key for both short and long-term food security, including the price stability. In this context, the land resource endowment and agricultural practices become crucial, as resource endowment of each country determine their priorities.

The existence of Bangladesh in 1971 happened when Bengali East Pakistan seceded from its union with West Pakistan. Located in Southern Asia. Bangladesh is bordered between Myanmar and

India with Bay of Bengal on its South. Bangladesh is an impoverished overpopulated country with a population of 141,340,476 (July 2004 est.) and a total area of 144,000 sq. km., situated on one of the most flat alluvial plain in the world. Land use in Bangladesh (2001) depicts that despite having an arable land of about 62.11 per cent, the food self-sufficiency has been fluctuating, even during the post-green revolution phase. It is primarily due to this reason that agricultural land use, productivity and the aspect of food self-sufficiency is questioned as a matter of serious introspection in Bangladesh.

The size and composition of the existing population and its growth over the period have an impact on various aspects of development including food security. Food supply is the basic element of food security. Food security forms the basis for social and economic security and hence socio-economic development. The ultimate goal of better quality of life depends on population, resources and environment. These factors are closely interrelated and interdependent. During the last few decades, the world as a whole experienced tremendous demographic upsurge, which is often termed as 'population explosion', and it occurred mostly in the poorer parts of the world.[2] A rapid decline in mortality and slower decline in fertility rates have been responsible for the rapid population growth in the developing world during the last four decades. Since the population was growing rapidly in the poorer parts of the world, food requirements were also pressing seriously. During the 1970s and 1980s, the food situation improved and the optimism about the food situation was revived.[3] In 64 out of 105 developing countries, food production lagged behind population growth over the period 1985-95.[4]

The late 1980s had witnessed a fall in per capita food production in developed countries and a slowdown in the rate of growth of production and yield in the developing countries. The problem since then is not merely been shortage of food production or availability but has been related with the purchasing capacity of the lowest income group. In developing countries rapid population growth has added burden on countries to keep pace with the production of foodgrains to make the food available as well as improving the buying power of the poor. Majority of population in Bangladesh is rural. Its growth is faster than many Asian countries.

**Agricultural Growth in Bangladesh**

Unarguably, food self-sufficiency is prerequisite in the resource starved and densely populated region. It is a matter of serious concern, since its natural resource base is further shrinking and the post-green revolution symptoms are not encouraging.[5] Several countries have already neared the limits of cultivable land, an essential component for any assessment and major determinant of present and future food production. In densely populated developing countries like China, Egypt and Bangladesh, the total farm area is shrinking, as good farmland is concreted over for roads, housing and factories.[6] Apart from inequality in distribution of wealth, the shortage of agricultural land is another cause for hunger and under-nutrition in Bangladesh imposing limits on adequate food production contributing to the economic causes of hunger. Deltaic Bangladesh though naturally fertile country has other natural disasters like cyclones and floods, which aggravate hunger. On the national level the lacunas of agricultural policies and projects have caused faltering food production in Bangladesh with market-oriented (profit-making) systems replacing the traditional food production systems which aimed at food self-sufficiency. However, examining food self-sufficiency becomes an imperative in terms of the pressure on land resources in Bangladesh due to increasing population. In the context of South Asia, Bangladesh is a very densely populated country with agrarian economic background. Therefore, the question of agricultural self-sufficiency holds the key vis-à-vis food productivity and population growth. The steady population growth in Bangladesh should also be looked into the Islamic nuances of birth control, where despite Bangladesh being an Islamic country, the birth control measures have been seriously undertaken. Heavy investment oriented solutions like education to control the escalating growth rate of population in Bangladesh are also being incorporated in policies. However, population cannot be completely held responsible for straining food production system, as it is also an attributer of increasing the possibility of food production efficiency.

Bangladesh has been historically popular regarding its productive agriculture, fine textiles and handicrafts in general. This pre-colonial prosperity of Bangladesh (then undivided India) was due to high agricultural productivity. Historic accounts dating as early as 1600 describes Bangladesh as an agriculturally abundant and agricultural prosperous land. Export commodities included finest cotton, silk, rice, butter and livestock to countries like Persia, Mesopotamia, Arabia, Europe and Japan. The British colonial legacy destroyed the economic framework of Bengal systematically. The initial destruction of the cotton textile industry followed by the agricultural destruction crippled the economy of Bengal. The British strategy of expanding the area under cash-crops such as tobacco, cotton, sugarcane, indigo and jute had negative impact. This alteration not only damaged the agricultural system but also taxed the soil and accelerated soil-erosion. The introduction of zamindari system by the British destroyed the incentive of increased food-production and made them dependent on industrial countries on food. The British hence interfered with Bangladesh agricultural system by misusing the land, introducing unsuitable land tenure system.

Apart from colonial exploitation and alteration of agricultural system, Bangladesh witnessed further agricultural exploitation even as East Pakistan. This caused and escalated the debacle of hunger and poverty in 1950s and 1960s. Further, the misplaced economic policy was of encouraging industrial development of West Pakistan at its expense. This discriminatory economic policy of West Pakistan caused under-development apart from poverty spilling much in 1970s.

Post-1970s, the war in Bangladesh consequently crippled its economy further apart from the destruction of infrastructure. The development process hence started from a scratch and the new conundrum of hunger emerged. It is the high inequality in the distribution of income and land resources apart from inadequate employment opportunities which led to the escalation of hunger in Bangladesh. Also the vicious circle of a natural calamity like a flood, drought, crop diseases and cyclone leads to food shortage situation which leads to food hoarding syndrome by big farmers for raising food prices and creating artificial scarcity.

### *Demographic Trend*

For the last 50 years, the mushrooming growth of population and the path of future growth have created both alarm and confusion about the demographic changes in the South Asian region and especially in Bangladesh. In all the countries of South Asia population has been increasing quite significantly since 1950.[7] With the decline in mortality rate due to improved medical facilities and almost unchecked fertility in the coming years, it is going to exceed the East Asian region including China, by 2025.[8]

Agrarian development in Bangladesh has taken place against a backdrop of rapid population growth and *vice-versa*. The expansion of agriculture in under-utilised regions during the colonial rule gave enough scope to population expansion. Further, due to an immature manufacturing sector the pressure on agrarian land increased. According to Blyn[9] (1966), the population density in Bengal around the middle of the 18th century was twice higher than Punjab. The population of Bangladesh was around 14.5 million in 1801. It had doubled after 100 years in 1901, to 28.9 million.[10] The estimated growth rate was almost constant at a level of 0.67 per cent. The slow growth of population continued till 1931 — only 0.74 per cent of exponential growth rate. The doubling time of the population of 1931 was only 40 years. As shown in Table 1 at the inception of Bangladesh, it was 76.4 million showing the highest growth rate. The alarming demographic pressure generated by the growth of population has crippled all other efforts of development. Table 1 gives the lastest picture of population trend in Bangladesh.

**Table 1:** Population Growth in Bangladesh

| *Census (Years)* | *Population (in Million)* |
|---|---|
| 1901 | 28.9 |
| 1911 | 31.5 |
| 1921 | 33.2 |
| 1931 | 35.6 |
| 1941 | 42.0 |
| 1951 | 44.1 |
| 1961 | 55.2 |
| 1974 | 76.4 |
| 1981 | 89.9 |
| 1991 | 109.9 |

*Source*: Bangladesh Bureau of Statistics 1995, Dhaka.

Population growth of my country is directly influenced by its demographic characteristics, and Bangladesh is no exception. Till 1990, there was hardly any sign of decline in the birth rate. It started declining by the end of 20th century.

According to the World Development Report,[11] Bangladesh is expected to reach the net reproduction rate of which will bring fertility below replacement level by the year 2010. Although population growth rate has declined, the population momentum is continuing, which will necessitate more food grain production.

## Agricultural Land Use, Performance in 1980s and 1990s and Self-sufficiency

In one of the most densely populated countries of the world, agricultural land has reached its limit there exists no scope for the expansion of such land. During 1990-99, the increase of agricultural land was negative in Bangladesh. The high population density has led to a reduction is agricultural land.

Land scarcity is being recognised as a major problem. It affects not just country's overall ability to feed its population, translated down to individual and local level, if it is not matched step by step by rising yields, it means rising poverty and landlessness.[12] The agricultural growth of Bangladesh, strictly adheres to the above statement. Since the term "Green Revolution" was coined and it came as an effective force erasing the "begging bowl" image. The main features of the foodgrain sector in Bangladesh emerged in the 1970s and 1980s. The Green Revolution, which started from India, hardly took anytime to change the balance sheet of Bangladesh. The foodgrain situation has improved considerably, despite the fact that the country depends heavily on the monsoon and suffers from chronic floods, cyclones and droughts. The effective utilisation of agricultural land played a major role in this improved balance sheet (Fig. 1). This was combined with advancement of irrigation facilities and subsidies to farmers. The increasing utilisation of modern inputs and latest technology in agriculture, to raise food production, can be treated as a response to the growing needs of increasing population.[13] Though, growth of cereal production in the last few years has been just short of the demographic growth rates.

So far the desired growth rate is concerned, the World Food Conference set a target of 4 per cent annual rate of food and agricultural production in less developed countries like those of South Asia.[14] The performance of all the countries fell below target and Bangladesh is no exception [Table 1(a)]. Here, we can see that the projected and estimated growth rate is not that encouraging.

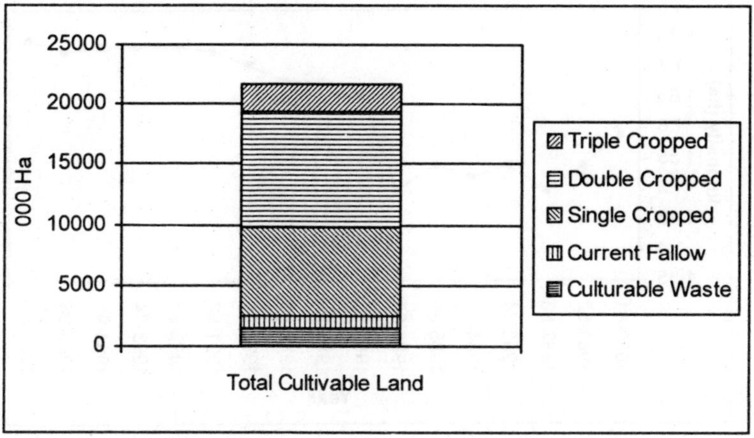


*Source*: BBS Yearbook 1996.

**Fig. 1:** Distribution of Cultivable Land 1995

**Table 1(a):** Average Annual Agricultural Growth Rate.

| *1970-80* | *1980-92* | *1992-2000* |
|---|---|---|
| 2.6 | 2.3 | 1.8 |

Though the scarcity of agricultural land was compensated by high agricultural yield during 1980s and till mid-1990s, the stagnating yield has once again brought the question of food self-sufficiency back in the light of liberalisation, privatisation and globalisation. The declining agricultural growth is quite significant considering the fact that agriculture is the most important sector of Bangladesh's economy. It is the principal source of income and livelihood of the vast majority of the population, employing approximately 66 per cent of the labour force. Fifty-seven per cent of the labour force is engaged in the crop sector, which represents about 78 per cent of the value added in the agricultural sector.[15] Within the crop sector, rice dominates with an average of 71 per cent share of the gross output value of all crops. As a result, growth in the agricultural sector in the recent past is actually growth in rice production. Improved yield played a major role in increasing the overall production of rice and wheat (Fig. 2).

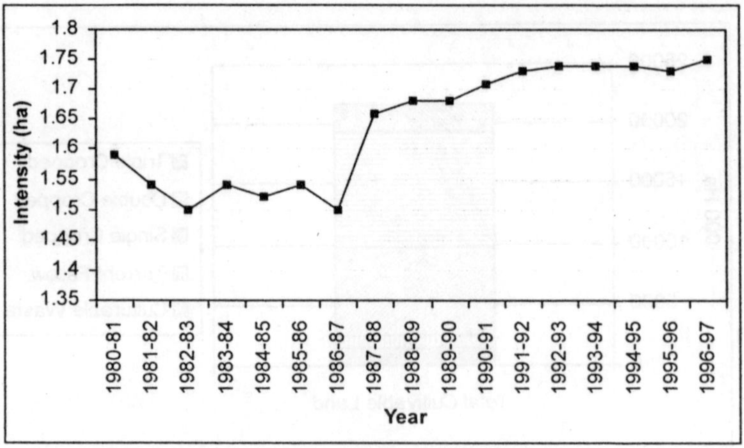


**Fig. 2:** Cropping Intensity (1980-97)

## Cropping Intensity

It is obvious from the above discussion that the increased production of foodgrain, which was necessary to meet the growing requirements, relied heavily on the success of Green Revolution. In the light of fragmented landholdings, owing it to culture and tradition, the success achieved is tremendous. This indicates that agriculture became more intensive. Double and triple cropping became norm in most areas. The figure given below sums up the cropping intensity between 1980-81 and 1996-97. It clearly appears that the agricultural intensification is being pursued vigorously and it became more intense after 1989-90, to an extent that on an average, on every ha. of net sown area, around two crops are being taken. This intensification was partly made possible after the introduction of irrigation. As a result of a sharp increase in cultivation of irrigated land during *rabi* season, which has resulted from the expansion of irrigation facilities, particularly systems served by tubewells, the cropping intensity increased.[16] Micro-level variability in agro-climatic conditions has fostered the emergence of a large number of localised cropping patterns, dominated by cereals. In most areas, the cropping pattern revolves around one, two, and sometimes even three rice crops, reflecting the paramount importance of rice in the rural economy and cropping intensity.[17] The domination of rice reflects from the fact that it accounts for 75 per cent of the total area (Fig. 3).

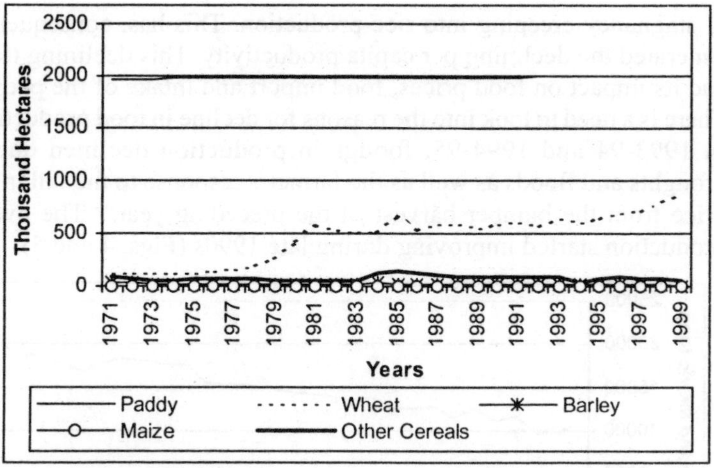


*Source*: http://www.sdnbd.org

**Fig. 3:** Area Under Paddy and other Crops (1971-1999)

### *Rice Production*

The mainstay objective of all agricultural and development policies and development since independence in Bangladesh has been to achieve self-sufficiency in foodgrains and in particular, rice production. Actually, rice is the principal staple food of Bangladesh and constitutes 95 per cent of the cereals consumed. It accounts for more than 74 per cent of cultivated area, 83 per cent of all irrigated area and 88 per cent of the total fertilizer consumption in the country.[18]

So far its significance in dietary habit is concerned, it accounts for 68 per cent of caloric intake and 54 per cent of protein intake. Socially, politically and economically, this is the most significant crop in Bangladesh.

Since its independence, Bangladesh is looking forward to achieve what is termed as 'self-sufficiency' in rice production. For this Bangladesh has been trying hard. It has been aiming all along since her First Five Year Plan (1973-78). To realise the importance of breaking the virtual stagnation in grain and agricultural production and attaining self-sufficiency at an initial stage. The Second Five Year Plan (1980-85) came within the slogan of "food production beyond self-sufficiency in the shortest possible time as its primary objective". This shows declining trend with a good beginning in 1989-90. Time-series annual data of rice production clearly shows (Table 2) the kind of stagnancy creeping into rice production. This has, consequently generated the declining per capita productivity. This declining trend had its impact on food prices, food import and intake of the people. There is a need to look into the reasons for decline in food production. In 1993-94 and 1994-95, foodgrain production declined due to droughts and floods as well as the farmer's response to the fall in the price from the bumper harvest of the preceding year.[19] The overal production started improving during late 1990s (Figs. 4 and 5).

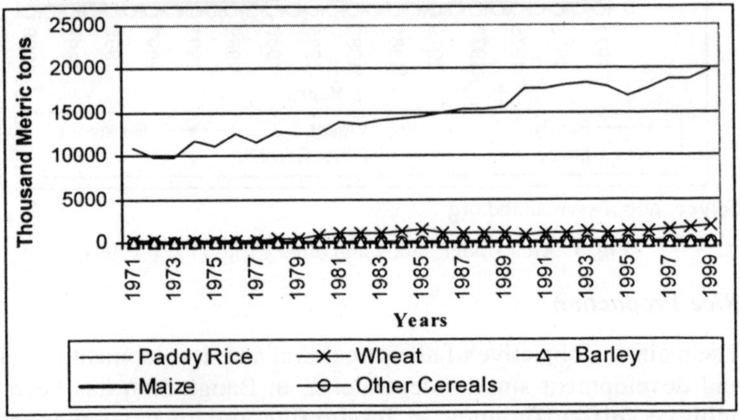


*Source*: http://www.sdnbd.org/sdi/issues/agriculture/database/production.htm

**Fig. 4:** Production of Paddy and Other Crops: 1971-1999

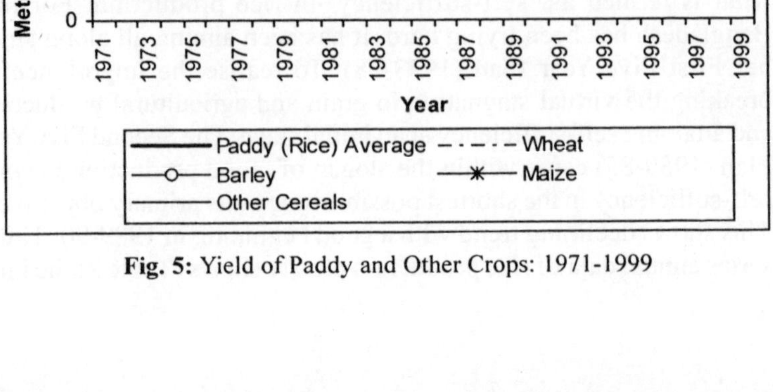


**Fig. 5:** Yield of Paddy and Other Crops: 1971-1999

**Table 2:** Per Capita Production of Rice in Bangladesh.

| *Year* | *Total Production of Rice (000' MT)* | *Net Production of Rice (000' MT)* | *Population (million)* | *Per Capita Production (Kg.)* |
|---|---|---|---|---|
| 1984-85 | 14689 | 13320 | 99.2 | 133 |
| 1985-86 | 15040 | 13536 | 100.7 | 134 |
| 1986-87 | 15407 | 13866 | 101.5 | 137 |
| 1987-88 | 15413 | 13872 | 103.4 | 134 |
| 1988-89 | 15544 | 13989 | 105.5 | 133 |
| 1989-90 | 17856 | 13071 | 107.5 | 149 |
| 1990-91 | 17852 | 16067 | 109.6 | 147 |
| 1991-92 | 18252 | 16427 | 111.7 | 147 |
| 1992-93 | 18342 | 16508 | 113.8 | 145 |
| 1993-94 | 18041 | 16237 | 116.3 | 140 |
| 1994-95 | 16832 | 15149 | 118.7 | 128 |
| 1995-96 | 17691 | 15922 | 121.1 | 132 |

*Note*: Production is netted by 10% for seed, feed and wastage.
*Source*: BBS 1996 Bangladesh Bureau of Statistics, 1996.

Foodgrain balance sheet explains the food situation in the mid-nineties. The per capita foodgrain availability in Bangladesh is on the declining trend. This indicates that the country's self-sufficiency is once again decreasing and people are still threatened by chronic food insecurity. The annual data for one decade suggests that there is a need to increase food production to have food availability at the national level which is prerequisite for food security. To make the declining trend more obvious the annual food gap for a decade is shown in Table 3.

In the mid-eighties when the green revolution was picking up, the gap seemed to be bridging. Only 1988-89 witnessed the widest gap in a decade of 3 million tons. This situation declined in the mid-nineties.

Basically, the agricultural sector is confronted with landholding size, pressure on limited land, low and stagnating yields of most crops, including rice, and the food gap between domestic production and demand has widened.

**Table 3:** Food Gap in Bangladesh.

| *Year* | *Requirement (million tonnes)* | *Net Production (million tonnes)* | *Gap (million tonnes)* |
|---|---|---|---|
| 1984-85 | 16.0 | 14.5 | 1.5 |
| 1985-86 | 16.5 | 14.4 | 2.2 |
| 1986-87 | 17.0 | 14.9 | 2.1 |
| 1987-88 | 17.5 | 14.9 | 2.6 |
| 1988-89 | 18.0 | 15.0 | 3.0 |
| 1989-90 | 18.4 | 17.0 | 1.4 |
| 1990-91 | 18.7 | 17.1 | 1.6 |
| 1991-92 | 18.9 | 17.5 | 1.4 |
| 1992-93 | 19.1 | 17.5 | 1.6 |
| 1993-94 | 19.5 | 17.3 | 2.2 |
| 1994-95 | 19.9 | 16.3 | 2.6 |

*Source*: FAO, the state of Food and Agriculture, 1996.

## Significance of Non-Rice Crop Production

The structure of crop production in Bangladesh has undergone changes which influences the food and nutritional security of the country. Although the "Green Revolution" was restricted to 'rice' in the initial stages, but later on 'wheat' got importance in the drier areas. It has recorded impressive increase in area and production. In a rice dominated food habit it was very difficult to convince people to have wheat or grow wheat. Actually, it was the food crisis of 1973, which forced the import of wheat bringing in some change. The yield per hectare of most crops increased in the last two decades, but the increase was most spectacular for wheat. The total area under wheat also increased. Since rice has already been stretched to its ecological limits because ecological requirements for rice are quite demanding, wheat gave some hope of bridging the production and demand. In many regions of Bangladesh, wheat is an attractive crop. Although *boro* rice tends to be more profitable in financial terms, it cannot be grown everywhere. For many rural households, the decision to grow wheat is influenced by the desire to avoid seasonal food shortages.[20]

Another important crop which is the main source of nutrition is pulse. Here, pulses are the main source of protein. Pulses, being leguminous crops, are important source of supplying nitrogen to the soils. Therefore, pulses are not only important in influencing the nutritional status of the people but also agricultural productivity. But its production and area did not increase very significantly. The total area under pulses increased steadily between 1971 and 1999, from 3,34,000 ha. to 5,53,800 ha.

From the above discussion one can derive that (a) cereal production has gone into a stagnation (b) production of non-rice crops like pulses are non-increasing, thereby, making people more vulnerable to nutritionally insecurity. Therefore, bridging food gap is the need of the hour. By the year 2020, food production could fail to keep pace with increase in the demand for food by the growing populations in many developing countries.[21] Bangladesh, where all arable land has been brought under cultivation has to think of its population pressure and agricultural productivity.

## Conclusion

Facing tough challenges year after year, witnessing the onslaught of nature and rise in human number, a remarkable agricultural transformation has been achieved by Bangladesh in the last 25 years. The sustained efforts from the government, international agencies and non-governmental organisation, saw a complete change in its image from huge importer of foodgrain to almost self-reliant in major staple crop. Despite all these agricultural constraints such as "sever scarcity of arable land, the large and fast growing population, concentration of land in few hands and high frequency of natural disasters, Bangladesh has managed to bring down the incidence of hunger and food poverty considerably.[22]

The success was achieved due to increase in cropping intensity and increase in productivity, leading the nation across her medium-term goal. Now the policy-makers will have to tread a tougher rout, i.e. sustaining the momentum. Having realized its potentials and constraints, the increase in the productivity of rice will have to be maintained for self-sufficiency, without ignoring the diversification. Maintaining productivity growth across the entire agricultural sector will depend on increasing productivity among secondary crops and 'niche' commodities that exploit specific locational and seasonal advantages. The agricultural land use did not change completely, except the introduction have mostly influenced agricultural land use in Bangladesh.

## NOTES

1. Food security as "physical, social and economic access to food by all people at all times to sufficient, safe and nutritious food to meet their dietary needs according to food preferences for an active and healthy life".
2. FAO, *Food Requirements and Population Growth*, WFS Tech/10, 1996.
3. N. Islam (ed.) *Population and Food in the early twenty-first century. Meeting future food demand of an increasing population"* IPFRI, 2020 vision 1995.
4. *Population and Sustainable Development-Five Years after Rio*, UNFPA, 1998.
5. Mrityunjay M. Jha, "*Food Security-Dynamics and Dimensions*", 2003. p. 11.
6. *Ibid*. p. 15.
7. *Population Bulletin*, Vol. 49, No. 1, 1994.
8. Blyn (1966) *Agricultural Trends in India: Output, Availability and Productizvity*, University of Pennsylvania Prem, Philadelphia.
9. M.A. Islam *Population Policy: Thinking Ahead A Review of Bangladesh's Development 1995* pp. 292, centre for policy Dialogue, Dhakha.
10. *World Development Report*, 1994.
11. *World Deveopment Report*, 1994.
12. *Ibid.,* p. 16.
13. Mrityunjay M. Jha, *op. cit*, p. 36.
14. B.M.Bhatia, *Food Security in South Asia*, Oxford and IBH Publication Company, New Delhi, 1985.
15. FAO (1996), *The State of Food and Agriculture,* p. 131.
16. Michael L. Morris, Nuimuddin Chodhury and Craig Meisner (1997), "*Wheat Production In Bangladesh-Technological, Economic and Policy Issues*", Research Report 106, IFPRI. p. 10.
17. *Ibid.*
18. M.M. Rashid (1994) *Achievements, Constraints and Future Activities in Rice Research and Production in Bangladesh,* Bangladesh Rice Research Institute. In FAO (1996), the State of Food and Agriculture.
19. Government of Bangladesh/FAO (1996), *Strategies and programme Framework for Agricultural development in Bangladesh.*
20. Michael L. Morris, Nuimuddin Chodhury and Craig Meisner (1997), "*Wheat Production In Bangladesh-Technological, Economic and Policy Issues*", Research Report 106, IFPRI. p. x.
21. IFPRI (1997), Food Gap Widening in Developing Countries, *IFPRI News Release*, 26 October.
22. P.L. Joshi, "The Demography of South Asia" in Ashish Bose and M.K. Premi (ed.) *Population Transition in South Asia.*

## REFERENCES

Bhatia, B.M. (1985): *Food Security in South Asia*, Oxford and IBH Publication Company, New Delhi.

Boserup, E. (1965): *Conditions of Agricultural Growth: The Economics of Agrarian Change Under Population Pressure.* Aldine, New York.

Centre for Policy Dialogue, (1996): Growth or Stagnation: *A Review of Bangladesh Development 1996,* Centre of Policy Dialogue, Dhaka.

FAO. (1996): *Food Requirements and Population Growth,* WFS Tech/10. FAO, Rome.

Islam, N. (1995): *Population and Food in the early twenty first century. Meeting future food demand of an increasing population,* IPFRI, 2020 Vision. Washington D.C.

FAO. (1996): *The State of Food and Agriculture.* Rome.

Government of Bangladesh/FAO (1996): *Strategies and Programme Framework for Agricultural Development in Bangladesh.* FAO, Rome.

Hossain, M. (1977): *Agrarian Structure and Land Productivity in Bangladesh,* Cambridge University, Cambridge.

Hossain, M. (1991): *Agriculture in Bangladesh—Performance, Problems and Prospects.* University Press Limited, Dhaka.

IFPRI. (1997): *Food Gap Widening in Developing Countries,* IFPRI News Release, 26 October.

Islam, M.A. (1995): *Population Policy: Thinking Ahead A Review of Bangladesh's Development,* pp. 292, Centre for Policy Dialogue, Dhaka.

Islam, N. (1991): Bangladesh Agriculture: Growth Stability, and Poverty Alleviation. *Journal of International Development,* Vol.3, No. 5, pp. 447-465.

Jha, Mrityunjay M. (2003): *Food Security-Dynamics and Dimensions.* Northern Book Centre, New Delhi.

Morris, Michael L. *et al.* (1997): *Wheat Production In Bangladesh—Technological, Economic and Policy Issues,* Research Report 106, IFPRI.

Murshid, K.A.S. (1987): Instability in Foodgrain Production in Bangladesh: Nature, Levels and Trends. *The Bangladesh Development Studies,* Vol. 15, No. 1, pp. 33-73.

Rahman, Atiq and Hossain, M. (1985): *Agrarian Reform and Rural Poverty Alleviation—A Review of Experiences of Some Asian Countries.* FAO, Rome.

UNFPA, *Population and Sustainable Development—Five Years After Rio,* UNFPA, 1998.

Rahman, Atiq. (1987): Resource Use in Bangladesh Agriculture—An Empirical Analysis. *The Bangladesh Development Studies,* Vol. XV, No. 2. *Population Bulletin.* 1994. Vol. 49, No. 1.

Timmer, C.P. (2000): The Macro Dimensions of Food Security: Economic Growth, Equitable Distribution, and Food Price Stability. *Food Policy,* Vol. 25, pp. 283-295.

*World Development Report,* (1994).

*World Development report,* (2004).

http://www.sdnbd.org/sdi/issues/agriculture/database/area.htm

# 7

# Agricultural Transformation and Crop Combination Regions in Himachal Pradesh

KAUSHAL K. SHARMA

## Introduction—Dynamics of Agricultural Transformation

Factors affecting agriculture are dynamic. If out of physical, economic and social factors only one is taken to explain the spatial organization of agriculture, it becomes difficult to judge the influence, because it is very rare to come across an instance of a single factor controlling the distribution of agricultural activities. At the same time agriculture pattern is the result of interaction among the influence exerted by the physical, economic and social factors. Agricultural activities are primarily influence by the combination of natural environmental condition and human circumstances. While geographers have tended to emphasise the role of physical environment (Munton, 1969), economists on the other hand have laid overwhelming emphasis on economic factors in conditioning the distribution of geographic phenomena, whereby the mutual relationship between the phenomena has been over emphasized (Chisholm, 1966). However, one cannot ignore the impact of the vast multiplicity of interrelated physical and non-physical factors contributing to agricultural growth. However, one cannot ignore the impact of the vast multiplicity of interrelated physical and non-physical factors on agriculture, notwithstanding the fact that all of them are not equally significant in influencing regional variations and temporal development of agricultural phenomena in an area. Hence, agriculture is also dynamic when comparisons are drawn with old generations and old methods.

The major centre on farmer's actions is the physical environment which again does not act in under deterministic manner but rather lays down broad controls over what can and cannot be grown and over yields which can be expected for the given labour, fertilizers and other factors of production. Economic and physical factors interact with each other in a complex manner and together establish a range of possible forms of production, one or more of which may be chosen. The recognition of set of economic and physical controls implies that each farmer has best solution to the problem of what agricultural products he should produce, given his environmental and economic circumstances.

There is differential rate of agriculture production when comparisons are drawn between old generation and old methods and new generation and new methods. The new methods give rise in the scale of production and profits in agricultural activity. In the developed countries the agriculture sector is small and the share in economic growth is low as compared to industry. Whereas in the developing countries, the agriculture sector is large but *contribution to economic growth is low*. Hence, there is stagnation in agricultural development, because farmer does not save enough to reinvest in agriculture. Therefore, the supply of Capital from agriculture is low and contribution of family labour increases in agriculture sector. Agriculture being a source of economic growth, now it is an uphill task to see that how cheaply and how much growth can be realised from transforming traditional agriculture, by means of investment into more productive sector. Basically transformation is dependent upon investing in agriculture.[1]

World agriculture, in fact, comprises two distinct types of farming: (1) The highly efficient agriculture of the developed countries, where the substantial productive capacity and high output per worker permits a very small number of farmers to feed entire nation; and (2) The insufficient and low productivity agriculture of developing countries, where in many instances the agriculture sector can barely sustain the growing population, let alone the burgeoning urban population, even at a minimum level of subsistence.[2]

In developed countries, there has been steady growth of agricultural output since the mid-18th century. This growth has been due to technological and biological improvements. It further accelerated after the First World War and particularly, after the Second World War. In 1820, American farmers were producing 4 times more than their consumption, in 1920 the same farmers produced enough for 8 persons and by 1987 they could feed 80 people.[3]

The picture is entirely different when we turn to the production experience of Third World Nations. In many countries, agricultural production methods have changed relatively slowly over time.

Agriculture is the main stay of low income countries. In these countries most of the farmers are still in subsistence class notwithstanding the tremendous advances in agricultural technology that have been made elsewhere in the course of present century. Subsistence agriculture is also traditional in consistency of the methods of production used and of the commodities produced.[4]

In India, social changes and economic needs of the people have greatly promoted the farmers to adjust and readjust their land use practices, nature of crops grown, number of harvest and quantum of production of cereals and other crops commensurate with changing pattern of agricultural land use in harmony with them. Since independence several agricultural innovation have been introduced in India, in order to increase agricultural production to become self-sufficient. The constant, increase in total food production was 177 million tonnes in 1988-91 despite numerous constraints, has been most spectacular achievement of Indian Agriculture during post-Independ-ence period.[5]

The agriculture and economy of Himachal Pradesh remained unchanged for a long historical epoch. The basic instrument of production remained light wooden plough, sometimes iron-tipped and sometimes not. Uncertainty of rainfall made cultivation dependent on irrigation—wherever possible by Khuls, Lift Irrigation, Tubewells etc. Cropping pattern closely followed the soil and climatic pattern prevailing in each particular region. The system, primitive as it might appear, is well suited to local condition and is by no means inefficient by standards of traditional agriculture, the seed yield ratio of wheat and other crops is more favourable than that of few better producing areas elsewhere in India.

In Himachal Pradesh agriculture still is primitive, characterized by ignorance of modern techniques and resultant low productivity of land and labour. The numerous constraints preventing modernization, viz. economic, social, political and institutional, tend to perpetuate the existing situation, despite the tremendous advances in agriculture technology in the recent decades. In the state it is seen that agriculture has been rapidly changed by the people to meet their desired goals and objectives. Moreover, agriculture has continuously adjusted to change as a consequence of overall development. As a consequence of the nature of agriculture the state has made a breakthrough by initially increasing its food production. At the same time it has carved out a place for increasing the area under fruits, seed potato, ginger, edible mushroom, top quality off-season vegetables and seeds of temperate type of vegetables.

Therefore, the present study has examined and evaluated the agricultural transformation in Himachal Pradesh, rise and fall of new agriculture communities, production technologies shifting government policies, etc. It has also brought out the transformation of traditional agriculture.

In order to find out the transformation, various paramounts were taken, viz. agriculture land, cropping pattern, intensity of cropping, and crop combination analysis. Agricultural productivity—based on prices and calories, to know the spatio-temporal variation from 1970-71 to 1988-89 with three years interval has been attempted. The most appropriate statistical technique to know the trend of transformation is based on exponential time series.[6] These results showing positive or negative trend of change in 1988-89 over 1970-71 are arranged into different categories. Trend value indicate the main forces responsible for boom or depression.

## Farming Calendar

Before analyzing the cropping pattern and growth levels in terms of area, production and productivity of selected crops, it would be worthwhile to preface this part with some basic ideas about crops, their sowing and harvesting seasons.

There are two main crop seasons in Himachal Pradesh, e.g. kharif or the season of summer crops, and the Rabi or the season of winter crops. Crops like sugarcane do not fit into this scheme as it is planted in March and harvested in December or January. Rice, maize, pulses, ginger, turmeric and small millets are important kharif crops. Rabi crops include wheat, gram, barley, oil seeds, tobacco, etc. Unpredictable and uncontrollable monsoon, as well as high temperature and high humidity during kharif season leads to crops being prone to diseases. Hence, contribution of kharif crops to the total agriculture is less as compared to rabi crops.

While the sowing and harvesting time are commonly the same as in other parts of north Western India, in the low lying areas of this state, in the higher region, the time schedules are entirely different. In the higher mountains, kharif sowing takes place from March to July and harvesting is done from September to November. The crop is also known as an autumn crop. Rabi is sown from mid-September to mid-December and reaped in April in the valleys and in early July higher up. In the cold areas, crops are rather sown and reaped together and so often there is little time left in between to prepare in the fields thoroughly. There are great variation in sowing and harvesting with change in altitude.

Since this study is based on food and non-food crops of kharif and rabi season, therefore, it would be worthwhile to examine the trend in gross cropped area, total food crop area and total non-food crops area.

## Change in Gross Cropped Area

This shows the total area put under cropping. The gross cropped area has increased in 24 Unions of Himachal Pradesh, covering about 46 per cent area. It is due to the efforts of planned development. In 19 tehsils nearly 31 per cent area has shown declined, trend, due to the increase in area under plantation crops and non-food crops.

**Table 1:** Trend of Change in Gross Cropped Area in 1988-89 over 1970-71.

| *Sl. No.* | *Trend of Change* | *Per cent Range* | *No. of Tehsils* | *Total Area (Hectare)* | *Per cent to Total Area* |
|---|---|---|---|---|---|
| 1. | High | > 5.00 | 2 | 93671 | 2.77 |
| 2. | Medium | 5.00-2.50 | 3 | 104329 | 3.09 |
| 3. | Low | 2.50-0 | 19 | 1352600 | 40.10 |
| 4. | Low | 0-(-)2.5 | 13 | 726131 | 21.53 |
| 5. | Medium | (-)2.5-(-)5:00 | 6 | 328431 | 9.73 |
| 6. | High | > (-)5.00 | – | – | – |
| 7. | No Change | | – | – | 23.15 |
| 8. | Data not available | | 8 | 780789 | 100.00 |

From Table 1 it can be derived that much of the change in positive side is witnessed in the low category. Same is applied to the negative trend of change where 21.53 per cent area to the total area is in low category.

In regional analysis the sub-tropical area the expansion of gross cropped area is mostly seen in Una, Amb, Badsar, Bhattyal, Sadar (B), Ghunomur and Arki. Nalagrah, Kausli, Solan and Hamirpur tehsil highlight declining trend.

Similarly, in sub-temperate region, increase in Gross Cropped Area is seen in Chamba, Jogindernagar, Sadar (M), Sundernagar, Chachyort Karsogi. Where as is Gross Cropped Area is declining in Sarkaghat, Kadaghat, Shimla, Soni and Theog.

In temperate wet region out of 12 tehsils, low decline trend has emerged in 4 tehsils, viz. Bharmaur, Baujar, Kumarsain and Rohni. In rest 8 tehsils low to medium positive trend of change is recorded.

In the temperate dry region 3 tehsils out of 9 tehsils recorded low to medium trend of positive change. And remaining 6 tehsils are in the low to medium rate of change.

## Change in Food Crops

The major share in the cropped area goes to food crops which were 96.5 per cent to the gross cropped area.[7] It is an indication that the farmers give more importance to the growing of food crops than the non-food crops. The trend of change in food crops is given below:

**Table 2:** Trend of Change in Total Food Crops in 1988-89 over 1970-71.

| *Sl. No.* | *Trend of Change* | *Per cent Range* | *No. of Tehsils* | *Total Area (Hectare)* | *Per cent to Total Area* |
|---|---|---|---|---|---|
| 1. | High | > 5.00 | 2 | 93671 | 2.77 |
| 2. | Medium | 5.00-2.50 | 3 | 203601 | 6.03 |
| 3. | Low | 2.50-0 | 19 | 1248924 | 37.03 |
| 4. | Low | 0-(-)2.5 | 13 | 701923 | 20.81 |
| 5. | Medium | (-)2.5-(-) 5.00 | 6 | 399814 | 11.85 |
| 6. | High | > (-)5.00 | – | – | – |
| 7. | Data not available | | 8 | 780789 | 23.15 |

The rate of change over the period in food crops is not very encouraging. Nearly, 9 per cent area has shown medium to high change in Arki, Badsar, Tobhal, Chopal, Spiti and Amb tehsils. About 17 tehsils record very less positive change carrying about 37.03 per cent area. In the negotiable trend of change 20 tehsils with 32.66 per cent area, show that there is a more towards specialization and change in cropping pattern. The change is still low to medium over the period of time. None of the tehsils have shown high rate of negative change.

## Change in Total Non-Food Crops

In the State only 3.5 per cent area is occupied by non-food crops. The non-food crops constitute rape, mustard, linseed, groundnut, seasum, tobacco, tea, cotton etc. Out of the total area under non-food crops. Bulk of the area 62.9 per cent is covered by total oil seeds leaving other non-food crops with 37.1 per cent. The trend of change is given in the Table 3 given below:

**Table 3:** Rate of Change in Total Non-Food Crops in 1988-89 over 1970-71

| *Sl. No.* | *Rate of Change* | *Per cent Change* | *No. of Tehsils* | *Total Area (Hectare)* | *Per cent of Total Area* |
|---|---|---|---|---|---|
| 1. | High | > 5.0 | 2 | 93671 | 4.90 |
| 2. | Medium | 5.0-2.5 | 4 | 203601 | 15.09 |
| 3. | Low | 2.5-0 | 17 | 1248924 | 24.04 |
| 4. | Low | 0-(-)2.5 | 14 | 701923 | 13.88 |
| 5. | Medium | (-)2.50-(-) 5.00 | 6 | 399814 | 5.78 |
| 6. | High | > (-)5.00 | – | – | 5.60 |
| 7. | Data not available | | 8 | 780789 | 30.75 |

It is very interesting to note from the above table that people of Himachal Pradesh do not find food crops as very advantageous in raising their standard of living and income. They have started growing commercial crops mainly the oil seed. It is reflected when 23 tehsils covering 44 per cent area showing positive trend of change. Wherever the negative change has occurred the area is best for growing fruits and vegetables. The agriculture policy in Himachal Pradesh is to encourage the farmers to grow high return crops.

## Transformation in Cropping Pattern of Himachal Pradesh

Cropping pattern in Himachal Pradesh has not changed significantly over time and the new technology of agriculture production has not made a noticeable dent. The State enjoys a comparative advantage in the production of plantation crops, off-season vegetables, fruits and flowers. The land use pattern, of which the cropping pattern forms a part, has always been a dynamic phenomena. The cropping pattern is in fact, a reflection of the interplay of the complex social, economic and physical factors. All these factors themselves keep changing, except the physical ones, which are comparatively static. Thus under the influence of dynamic socio-economic factors, the cropping pattern also goes on getting modified and transformed, so much so, that, sometimes it gets absolutely replaced after a sufficiently long span of time.[8]

The farmers exercise his judgement for a most favourable selection of crops from a number of possible combinations on the basis of his experience.[9] Similarly, the cropping pattern indicates the level of development and the economic prosperity of the state. Lastly, the cropping pattern has became an important tool for planning rational and balanced programme for agriculture.

With this view the present study is done to evaluate the present cropping pattern situation and emerging trend of transformation in the system.

## Change in Area under Food Crops in Kharif Season

The kharif season usually extends from mid-June to mid-October. In higher mountains, kharif sowing takes place from March to July and harvesting is done from September to November. The crop is also known as an autumn crop in such area. The trend of change in area in food crops can be seen in the table below:

**Table 4:** Rate of Change in Total Non-Food Crops in 1988-89 over 1970-71.

| *Sl. No.* | *Rate of Change* | *Per cent Change* | *No. of Tehsils* | *Total Area (Hectare)* | *Per cent of Total Area* |
|---|---|---|---|---|---|
| 1. | High | > 5.0 | 3 | 153310 | 6.13 |
| 2. | Medium | 5.0-2.5 | 3 | 239931 | 7.11 |
| 3. | Low | 2.5-0 | 20 | 1118547 | 33.16 |
| 4. | Low | 0-(-)2.5 | 15 | 950501 | 28.18 |
| 5. | Medium | (-)2.5-(-)5.0 | 1 | 55742 | 1.65 |
| 6. | High | > (-)5.00 | 1 | 21027 | 0.62 |
| 7. | Data not available | | 8 | 780789 | 23.15 |

It is evident from Table 4 that 26 tehsils have shown positive trend of change out of which 20 tehsils are in the low category and 3 each in medium and high, accounting for 46.40 per cent area.

Meanwhile 17 tehsils accounting for 30.45 per cent describe negative trend of change.

The sub-tropical area, which is most suitable for food crops show high trend of positive change in Arki, Badsar and Amb and 5 tehsils account for low rate of change. The tehsil of Solan, Hamirpur and Kamli show high, medium and low change respectively.

In sub-temperate region which is also very fertile for food crops have highlighted that the nearly 6 tehsils fall under change positive and 5 tehsils in negative rate of change. Similarly, it is repeated for the temperate wet region. In the temperate dry region 4 tehsils lie in positive side and 4 on the negative side. One trend has emerged that all the valley region and low lying area have shown positive rate of change. The higher altitude area have registered negative rate of change.

## Change in Area under Rice Cultivation

Rice is one of the most important crop grown in the State. It is the staple food of the people of Himachal Pradesh. Due to its water requirement, it is best grown in valleys and in area where irrigation is possible. It is mostly grown in sub-tropical and sub-humid region and in few valleys of temperate region. Looking at the amount of labour required to grow this crop most of the people in the state have shifted to raising of other crops which require less labour and provide higher income. Table 5 shows the rate of change in rice cultivation.

**Table 5:** Rate of Change in Area Under Rice in 1988-89 over 1970-71.

| *Sl. No.* | *Rate of Change* | *Per cent Change* | *No. of Tehsils* | *Total Area (Hectare)* | *Per cent of Total Area* |
|---|---|---|---|---|---|
| 1. | High | > 5.0 | 3 | 138624 | 4.11 |
| 2. | Medium | 5.0-2.50 | – | – | – |
| 3. | Low | 2.5-0 | 12 | 1055047 | 31.28 |
| 4. | Low | 0-(-)2.5 | 13 | 828246 | 24.56 |
| 5. | Medium | (-)2.5-(-)5.00 | 9 | 481874 | 14.28 |
| 6. | High | > (-)5.00 | 9 | 312182 | 9.25 |
| 7. | Data not available | | 5 | 558744 | 16.56 |

Table 5 highlights that positive rate of change is less in rice cultivation, whereas the non-traditional rice producing tehsils have started growing rice, viz. Ranpur, Rohuru, Chapal, Komarshir, Bhattiyat, Chichiyal and Karsig. The traditional rice producing tehsils have shown decline in rice cultivation. Most of the people with large size farms have shifted to other crop. But small farmers do not have option but to depend in the same crop.

The sub-tropical region which is considered to be the best for rice cultivation due to physio-climatic condition, highlights that out of 15 tehsils 4 have registered positive trend of change, rest all have negative change. In sub-humid region 4 tehsils have positive rate of change out of 14 tehsils. The same trend is seen in sub-humid region. The temperate region which is not suitable for rice cultivation have registered mostly a decline. The rice cultivation is confined to valley region.

### Change in Area under Maize Cultivation

Maize is another important crop of kharif season. It covers 30 per cent of grossed cropped area and its place in terms of percentage share to the total area is second in the state after wheat. The non-rice producing tehsils have more average of area under maize cultivation. In 1988-89 shows that 40 tehsils out of 51 tehsils have shown positive trend of change (Table 6).

**Table 6:** Rate of Change in Area Under Maize in 1988-89 over 1970-71.

| *Sl. No.* | *Rate of Change* | *Per cent Change* | *No. of Tehsils* | *Total Area (Hectare)* | *Per cent of Total Area* |
|---|---|---|---|---|---|
| 1. | High | > 5.0 | 8 | 570509 | 16.91 |
| 2. | Medium | 5.00-2.50 | 7 | 427647 | 12.68 |
| 3. | Low | 2.50-0 | 25 | 1418278 | 41.96 |
| 4. | Low | 0-(-)2.50 | 7 | 740972 | 21.97 |
| 5. | Medium | (-)2.50-(-)5.00 | 1 | 20016 | 0.59 |
| 6. | High | > (-)5.00 | – | – | – |
| 7. | Data not available | | 8 | 216864 | 6.43 |

Although, the table is self-explanatory to show that 77.55 per cent area comprising 40 tehsils account for positive rate of change, on the other hand, Bharmaur and Khagra valley and Pachhad valley in Sirmaur have registered negative trend of change. The whole Himachal has a good potential for Maize cultivation.

## Change in Common and Little Millets

In Himachal Pradesh, Common and Little Millets account for about two-third area under cultivation. It is mostly grown in the cold weather of higher valleys of Kinnaur, Lahaul and Spiti and Kullu districts. In the time period which is under observation shows that majority of the tehsils (24) have shown negative trend of change (Table 7).

**Table 7:** Change in Common and Little Millets in 1988-89 over 1970-71.

| *Sl. No.* | *Rate of Change* | *Per cent Change* | *No. of Tehsils* | *Total Area (Hectare)* | *Per cent of Total Area* |
|---|---|---|---|---|---|
| 1. | High | > 5.0 | 7 | 100421 | 2.97 |
| 2. | Medium | 5.0-2.5 | 2 | 189780 | 5.60 |
| 3. | Low | 2.5-0 | 3 | 151211 | 4.48 |
| 4. | Low | 0-(-)2.5 | 6 | 732114 | 21.70 |
| 5. | Medium | (-)2.5-(-)5.0 | 9 | 549427 | 16.29 |
| 6. | High | > (-)5.0 | 9 | 453800 | 13.45 |
| 7. | Data not available | | 15 | 1136508 | 33.70 |

From the above Table 7 it is clear that 12 tehsils accounting for nearly 13 per cent area have shown positive rate of change. Nearly double the figure of tehsils (24) from the above category show declining trend of change. Adding 15 tehsils to it, making a total of 39 tehsils highlight that people are not very interested to grow this crop. In the sub-tropical region it is not very well preferred crop, but wherever is grown for example Ponta and Pachhad the trend of decline have come into existence. Similarly, in the case of sub-humid region all the tehsils (16) have shown low to high rate of decline. In the temperate region, leaving aside the 4 tehsils of Kullu, 2 tehsils of Shimla district, rest all tehsils have shown declining trend. Same declining trend is seen in temperate region, which is considered to be the area for common Millet cultivation. With this analysis one thing surely comes out is that growing coarse crops are showing high transformation from higher category to lower category. Most of these are being replaced by fruit and vegetable crops.

## Change in Horsegram Cultivation

It is one of the most important pulses of the kharif season. It is mostly grown in sub-tropical temperate wet region comprising of 34 tehsils in the State. 24 tehsils has shown decline in the area under horsegram cultivation (Table 8).

**Table 8:** Change in Common and Little Millets in 1988-89 over 1970-71.

| *Sl. No.* | *Rate of Change* | *Per cent Change* | *No. of Tehsils* | *Total Area (Hectare)* | *Per cent of Total Area* |
|---|---|---|---|---|---|
| 1. | High | > 5.0 | 4 | 291617 | 8.64 |
| 2. | Medium | 5.0-2.5 | 2 | 163818 | 4.85 |
| 3. | Low | 2.5-0 | 4 | 251159 | 7.44 |
| 4. | Low | 0-(-)2.5 | 12 | 626517 | 18.57 |
| 5. | Medium | (-)2.5-(-)5.0 | 8 | 297498 | 8.82 |
| 6. | High | > (-)5.0 | 4 | 372438 | 11.04 |
| 7. | Data not available | | 17 | 1309397 | 39.40 |

The above Table 8 shows that very less positive trend of change is seen in the area under Horsegram cultivation (20.93%). Another 38.43 per cent has declined. The positive change is only in the river valley tehsils of the State. In 39.40 per cent area this is not grown.

The regional trend is not very well developed. In each region positive and negative trend is highlighted.

## Trend of Change in Total Pulses

Beside gram and masur all other pulses are grown in kharif season viz. Mash, mung, kulth and rong. According to 1985-86 figures nearly 4.5 per cent of the total gross cropped area is under the pulses. The pulses are mostly grown in Shimla, Kinnaur, Mandi districts of Himachal Pradesh. The trend of change is shown in the Table below:

**Table 9:** Trend of Change in Area under Pulses in 1988-89 over 1970-71.

| *Sl. No.* | *Rate of Change* | *Per cent Change* | *No. of Tehsils* | *Total Area (Hectare)* | *Per cent of Total Area* |
|---|---|---|---|---|---|
| 1. | High | > 5.0 | 3 | 356401 | 10.56 |
| 2. | Medium | 5.00-2.50 | 8 | 349240 | 11.69 |
| 3. | Low | 2.50-0 | 7 | 333623 | 9.89 |
| 4. | Low | 0-(-)2.50 | 13 | 760412 | 22.54 |
| 5. | Medium | (-)2.50- (-)5.00 | 6 | 298403 | 8.84 |
| 6. | High | > (-)5.00 | 8 | 824126 | 24.43 |
| 7. | No Change | | 1 | 82888 | 2.45 |
| 8. | Data not available | | 5 | 178433 | 5.29 |

In all 15 tehsils accounting for 32.14 per cent area depict positive trend of change. They form a region in the South-east of the State comprising Kinnaur (6 tehsils), Shimla (7 tehsils) and Pachhad (Armar). Out side this compact region Pangi, Narpur, Jogindernagar and Sundernagar tehsil medium to high rate of positive change is registered. The above Table 9 further highlights that 27 tehsils show declining trend. It implies that in Himachal Pradesh, the scope of growing pulses is declining.

## Change in Non-Food Crops in Kharif Season

In the state as a whole 3.5 per cent area came under non-food crops. The non-food crops contribute rape, mustard, linseed, seasum, tobacco, tea, cotton and shuhump, etc. Out of the total area under non-food bulk of the area, i.e. 62.9 per cent is covered by total oil seed leaving thereby 37.1 per cent share to other non-food crops.[10] For the present study seasum, mustard and condiments and spices have been included on the basis of high relative share of area under these crops. The area showing change is given in the Table below.

**Table 10:** Change in Area under Non-Food Crops in 1988-89 over 1970-71.

| *Sl. No.* | *Rate of Change* | *Per cent Change* | *No. of Tehsils* | *Total Area (Hectare)* | *Per cent of Total Area* |
|---|---|---|---|---|---|
| 1. | High | > 5.0 | 7 | 236453 | 7.01 |
| 2. | Medium | 5.00-2.50 | 2 | 40321 | 1.19 |
| 3. | Low | 2.50-0 | 5 | 414668 | 15.74 |
| 4. | Low | 0-(-)2.50 | 4 | 233957 | 6.93 |
| 5. | Medium | (-)2.50-(-)5.00 | 9 | 573260 | 16.99 |
| 6. | High | > (-)5.00 | 10 | 721528 | 21.39 |
| 7. | Data not available | | 14 | 1036982 | 30.75 |

In the above Table 10 the declining area is more than the increasing area, viz. 23.94 and 45.21 respectively. Rate of change in non-food do not form a very clear-cut regional pattern. The different agro-climatic regions form a cultivation of both positive and negative trend of change.

## Change in Mustard and Seasum Cultivation

They occupy a significant area under non-food crop. It is mostly cultivated in sub-tropical and sub-humid region in the state. Even in this regional context all tehsils do not grow mustard. Seasum is grown in 11 tehsils of the State. Out of which Palampur, Nalagrah and Kanli show very high negative rate of change. Meanwhile, Narpur, Amb, and Arki recorded high rate of change.

## Changes in Condiment and Spices

They include chillies, ginger, pepper, turmeric crops in this category. They provide high income (Cash) to the farmers and the amount of labour required is also less as compared to most of the food crops. From the table below majority of positive shift is on the increase from 1970-71 to 1988-89 (Table 11).

**Table 11:** Change in Area under Condiments and Spices in 1988-89 over 1970-71 in Himachal Pradesh.

| *Sl. No.* | *Rate of Change* | *Per cent Change* | *No. of Tehsils* | *Total Area (Hectare)* | *Per cent of Total Area* |
|---|---|---|---|---|---|
| 1. | High | > 5.0 | 14 | 836883 | 24.81 |
| 2. | Medium | 5.00-2.50 | 7 | 459346 | 13.62 |
| 3. | Low | 2.50-0 | 4 | 358264 | 10.62 |
| 4. | Low | 0-(-)2.50 | 4 | 342593 | 10.15 |
| 5. | Medium | (-)2.50-(-)5.00 | 6 | 386628 | 10.40 |
| 6. | High | > (-)5.00 | 5 | 299567 | 8.88 |
| 7. | Area not put to use | | 11 | 786055 | 23.30 |

Nearly 50 per cent tehsil (25 tehsils) covering 50 per cent area of the state show positive rate of change. 15 tehsils recorded negative rate of change with 29.43 per cent area. The most interesting feature that has emerged is that all the agro-climatic condition show low to high rate of change. In the outermost temperate zone the condiments and spices are not grown. In sub-tropical region, Ponta, Dehra and Narpur tehsils mark medium to high negative rate of change. Rest all tehsils record positive rate of change. In sub-humid zone Chamba, Kangra, Chachiyat, Suri and Kawaghat are showing negative rate of change. In the temperate wet region 6 out of 12 tehsils show decreasing trend.

## Change in Food Crops in Rabi Season

The rabi crop is sown from mid-September to mid-December and reaped in April in the valleys and in early July in higher slopes. There are great variations in sowing and harvesting with change in altitude. The rabi must be sown before the winter snowfall and is to be so timed that the seedling are not sufficiently grown up before the full onset of the winter season. Too early sowing will result in considerable growth which is liable to damage, causing delay in flowering and fruiting in the spring season after the snow melts. While early winter and spring frosts are harmful, snow cover is protective for the young seedlings during winter.[11] For the present study, wheat, barley, gram, masur have been evaluated to show the transformation in area under food crops. In the non-food category, the share of individual crop is so small to show any significant impact on the cultivation pattern, with respective all non-food crops in total have been included. The area under shift in food crops in rabi season is given below.

**Table 12:** Change in Area under Food Crops in Rabi Season in 1988-89 over 1970-71.

| *Sl. No.* | *Rate of Change* | *Per cent Change* | *No. of Tehsils* | *Total Area (Hectare)* | *Per cent of Total Area* |
|---|---|---|---|---|---|
| 1. | High | > 5.0 | 1 | 54268 | 1.60 |
| 2. | Medium | 5.00-2.50 | 4 | 181804 | 5.39 |
| 3. | Low | 2.50-0 | 20 | 11527184 | 34.18 |
| 4. | Low | 0-(-)2.50 | 13 | 1070906 | 31.75 |
| 5. | Medium | (-)2.50-(-)5.00 | 8 | 248144 | 7.35 |
| 6. | High | > (-)5.00 | 4 | 261265 | 7.74 |
| 7. | Data not available | | 1 | 39403 | 1.16 |

The above Table 12 shows that 25 tehsils lie in positive and negative rate of change. But the per cent of area under each category shows variation, viz. in positive change 41.7 area exists and in negative change 46.84 per cent had its existence. Only 1.6 per cent area lies in high category of positive change. Meanwhile, in the high negative change 7.74 per cent change is highlighted.

In sub-tropical region Dehra, Hamirpur, Nalagrah, Kaushi, Solan and Ponta show low to high negative change. High change is witnessed in Badsar followed by medium positive change in Ghumansin tehsil of Balaspur. Other tehsils touching Punjab and Haryana, viz. Phalligat, Nurpur, Amb, Una, Solan and Pachhad.

In sub-humid region the extension of positive and negative area of sub-tropical region is noticed. The Sru Kanga, and Kunihar valley in this region shows negative change covering 7 tehsils. Rest of 7 tehsils in higher location has shown positive trend of change. The temperate wet region is not a very high potential area for food crops. Wherever the positive change is registered it is mostly in the low category, viz. Churah, Bharmaur, Kullu, Anni, Rampur, Rolun and Tulshal. Very high negative change is in Nichar, Bangar and Katkahsi.

In the outmost zone (temperate-dry region) minor positive changes occur in Spiti and Songla tehsils. Rest all tehsils (7 tehsils) record negative rate of change.

### Change in Wheat Crop Area

Wheat ranks first among all the crops grown in the state. As 32 per cent area of the gross cropped area is under wheat. 48 per cent of the total cropped area is shared by wheat in Hamirpur district, followed by Bilaspur (43%), Kangra (41%), Mandi (40%), Una (40%). In the low category, Kinnaur account for 10 per cent area to total cropped area followed by Lahaul Spiti (18%). The tehsil-wise variation in the area under Wheat cultivation is given in Table 13.

**Table 13:** Change in Area under Wheat Cultivation in 1988-89 over 1970-71.

| *Sl. No.* | *Rate of Change* | *Per cent Change* | *No. of Tehsils* | *Total Area (Hectare)* | *Per cent of Total Area* |
|---|---|---|---|---|---|
| 1. | High | > 5.0 | 4 | 169664 | 5.03 |
| 2. | Medium | 5.00-2.50 | 5 | 436422 | 12.94 |
| 3. | Low | 2.50-0 | 22 | 1403604 | 41.62 |
| 4. | Low | 0-(-)2.50 | 13 | 870096 | 25.80 |
| 5. | Medium | (-)2.50-(-)5.00 | 6 | 247298 | 7.33 |
| 6. | High | > (-)5.00 | 1 | 195117 | 5.78 |

Nearly 60 per cent area has shown positive trend of change over the time period. This has been mainly due to specialization in this crop with regard to high yielding varieties of seeds and drought resistant seeds etc. About 40 per cent area of 20 tehsils has shown decline. In case of Kangra and Hamirpur districts as a whole, the area under Wheat is very high, but over the period they have shown a decline. Lahaul tehsil and Kinnaur district are not suitable for wheat growing as they record high to medium rate of negative change. The apple and grains producing in south-east and Citrus in the north-west have shown more declining trend.

## Change in Area under Barley Cultivation

It is another rabi crop particularly in the inner and higher parts of Himachal Pradesh. It is mostly grown in Lahaul Spiti, Kinnaur, Shimla, Kullu and in higher altitudes of Chamba and Mandi districts. Only 3.3 per cent area is under this crop. It makes a combination with wheat crop and it can adapt in temperate dry region have shown decline.

**Table 14:** Change in Area under Barley in 1988-89 over 1970-71.

| *Sl. No.* | *Rate of Change* | *Per cent Change* | *No. of Tehsils* | *Total Area (Hectare)* | *Per cent of Total Area* |
|---|---|---|---|---|---|
| 1. | High | > 5.0 | 2 | 108809 | 3.22 |
| 2. | Medium | 5.00-2.50 | 0 | – | – |
| 3. | Low | 2.50-0 | 15 | 956234 | 28.35 |
| 4. | Low | 0-(-)2.50 | 17 | 1014818 | 30.09 |
| 5. | Medium | (-)2.50-(-)5.00 | 8 | 869694 | 25.78 |
| 6. | High | > (-)5.00 | 7 | 340232 | 10.08 |
| 7. | Date not available | | 2 | 94549 | 2.80 |

Nearly (17 tehsils) Pradesh are showing positive change. In Himachal Pradesh, course grain crops are not favoured by the farmers. Hence, high rate of negative rate of change is seen with respect to this crop. 32 tehsils fall in temperate wet and dry region. Out of which 4 tehsils, viz. Spiti, Rolum, Kumarsain and Chapal have show medium positive rate of change rest all have recorded negative rate of change. The sub-tropical and sub-humid region the located higher altitude tehsils have marked positive rate of change.

## Change in Area under Gram Cultivation

Gram is one of the important pulse in the state, it is mostly grown in the sub-tropical and sub-humid region. It occupies 44.2 per cent area under cultivation in the state. The crop is grown mostly in 21 tehsils out of which 7 tehsils have registered positive trend and rest 14 tehsils highlight low to high rate of negative rate of change (Table 15). In rest 30 tehsils the crop is not grown.

**Table 15:** Change in Area under Gram in 1988-89 over 1970-71.

| *Sl. No.* | *Rate of Change* | *Per cent Change* | *No. of Tehsils* | *Total Area (Hectare)* | *Per cent of Total Area* |
|---|---|---|---|---|---|
| 1. | High | > 5.0 | 5 | 441173 | 13.08 |
| 2. | Medium | 5.00-2.50 | – | – | – |
| 3. | Low | 2.50-0 | 2 | 107544 | 3.18 |
| 4. | Low | 0-(-)2.50 | 1 | 59639 | 1.76 |
| 5. | Medium | (-)2.50-(-)5.00 | 1 | 206822 | 6.13 |
| 6. | High | > (-)5.00 | 12 | 719340 | 21.33 |
| 7. | Crop not cultivated | | 30 | 1837380 | 55.30 |

The regional pattern that emerges out from the analysis is that in temperate, wet and temperate dry regions the gram is not cultivated. In sub-humid region of Chamba and Shimla the grams do not have share in cropped area. Kangra, Palampur and Sundernagar tehsil recorded decline over the period of time. Jogindernagar, S. Ghat, Chachyot and Mdar (M) reflected high rate of positive change followed by Kazog and Kandraghat in low category. The sub-tropical region only Nurpur has positive trend rest all tehsils showing decline.

## Change in Area under Masur Cultivation

It is another important pulse grown almost in the same tehsils where the gram is cultivated. It is not grow in 32.50 per cent area comprising Lahaul Spiti, Kinnaur, Shimla and Sirmaur district in the State. Masur cultivation has shown positive trend of change over the time period from 1970-71 to 1988-89 (Table 16).

**Table 16:** Change in Area under Masur Cultivation in 1988-89 over 1970-71.

| *Sl. No.* | *Rate of Change* | *Per cent Change* | *No. of Tehsils* | *Total Area (Hectare)* | *Per cent of Total Area* |
|---|---|---|---|---|---|
| 1. | High | > 5.0 | 6 | 393651 | 11.67 |
| 2. | Medium | 5.00-2.50 | 7 | 348978 | 10.34 |
| 3. | Low | 2.50-0 | 4 | 273026 | 8.09 |
| 4. | Low | 0-(-)2.50 | 6 | 461496 | 13.68 |
| 5. | Medium | (-)2.50-(-)5.00 | – | – | – |
| 6. | High | > (-)5.00 | 7 | 689064 | 20.43 |
| 7. | Area not put to cultivation | | 21 | 1096059 | 32.50 |

The remaining 67.50 per cent area which is put under Masur cultivation, out of which 30.10 per cent mark positive change. Whereas, 37.40 per cent area has shown declining trend. The regional scenario do not highlight any definit trend or pattern. But in an average in sub-tropical region, the positive change is more than negative change. Same trend is applied to sub-humid region and wet temperate region.

## Change in Area under Total Pulses (Kharif)

The other pulses in the rabi season except gram and masur are grown in very small number. But they together account for more area and play a significant role in the total area put to food gram and pulses. The share of pulses has declined from 1970-71 to 1988-89 due to transformation of agriculture towards horticulture and specialization in Food-crops (Table 17).

**Table 17:** Change in Area Total Pulses in 1988-89 over 1970-71.

| *Sl. No.* | *Rate of Change* | *Per cent Change* | *No. of Tehsils* | *Total Area (Hectare)* | *Per cent of Total Area* |
|---|---|---|---|---|---|
| 1. | High | > 5.0 | 3 | 356401 | 10.56 |
| 2. | Medium | 5.00-2.50 | 8 | 394240 | 11.69 |
| 3. | Low | 2.50-0 | 7 | 333623 | 9.89 |
| 4. | Low | 0-(-)2.50 | 13 | 760412 | 22.54 |
| 5. | Medium | (-)2.50-(-)5.00 | 6 | 298403 | 8.84 |
| 6. | High | > (-)5.00 | 8 | 824126 | 24.43 |
| 7. | No change | | 1 | 82888 | 2.45 |
| 8. | Data not available | | 5 | 178433 | 5.29 |

The above Table 17 highlights that 55.81 per cent area shows transformation towards declining side. The position in the positive side is that 18 tehsils accounting 32.14 per cent lie in different locations in the State. Hence, for negative trend the sub-tropical region almost all leaving aside. Kausli and Solan tehsils have shown decrease in the area under pulses. Same is applied in sub-humid region, where only two tehsils have high positive change, viz. Palampur and Sundernagar rest of all tehsils have negative rate of change.

**Change in Area under Non-Food Crops in Rabi Season**

The area under non-food crops is very low as compared to food-crops in the state. As we have noticed earlier that there is decline in the area under food crops but at the same time the area is increasing under non-food crops. Various factors are responsible—both physical and non-physical. Recent effort of government in general and Horticulture and Agriculture Department in particular are responsible to transform old traditional agriculture into more paying commercial crops (Table 18).

**Table 18:** Change in Area under Non-Food Crops in Rabi Season in 1988-89 over 1970-71.

| *Sl. No.* | *Rate of Change* | *Per cent Change* | *No. of Tehsils* | *Total Area (Hectare)* | *Per cent of Total Area* |
|---|---|---|---|---|---|
| 1. | High | > 5.0 | 16 | 862292 | 26.56 |
| 2. | Medium | 5.00-2.50 | 7 | 507236 | 15.24 |
| 3. | Low | 2.50-0 | 8 | 593759 | 17.80 |
| 4. | Low | 0-(-)2.50 | 6 | 514746 | 15.36 |
| 5. | Medium | (-)2.50-(-)5.00 | 2 | 43316 | 2.28 |
| 6. | High | > (-)5.00 | 5 | 443844 | 13.46 |
| 7. | Data not available | | 7 | 295537 | 8.76 |

Nearly 60 per cent has positive trend of change and of which 26.56 per cent is the high range followed by low (17.80%) and medium (15.24%). In the negative trend 15.36 per cent area is under lower side and 13.46 per cent on the high side.

The regional pattern has very clearly emerged. The sub-tropical region three tehsils, viz. Sirmaur, Solan and Nalagrah have recorded low negative rate of change. Rest all tehsil in the high positive rate of change leaving aside Nurpur, Dehra and Hamirpur which are in medium category. Similarly, in sub-humid region 3 tehsils, viz. Kangra, Palampur and Thagur. Theog is in low rate mountain change category rest all are in high and low positive rate of change. The trend is again positive in temperate region with negative changes in Anni, Kumarsain and Kotlevai. The temperate dry (inter zone) is not suitable from pulses. In all tehsil of Kinnaur district no pulses are grown and the other 3 tehsils, viz. Pangi, Lahaul and Spiti negative rate is recorded.

## A Shifting Trend in Agriculture—A Crop-Combination Analysis

Agriculture is by far an important single use of land and the description of an analysis of various problems relating to agricultural activities have contributed a great deal to the contents of the regional and economic geography. Mostly, studies have, however, confined themselves to the description of dominant individual crops rather than complex agricultural systems, with a result that they usually suffer from over generalization, because the individual crops, except in relatively rare circumstances of extreme mono-culture, are not grown alone but characteristically they appear in combination (Weaver, 1954).[12] Weaver (1954) proposed the objective statistical technique of combinational analysis to establish crop, livestock and enterprise combinations involving both the qualitative and quantitative aspects of crop raising, livestock husbandry and functioning forms of agriculture, others.

Recently attention has been focused on combinational analysis of different element in geographical studies which give a scientific treatment of the geographical problem. Geographers are particularly interested in classifying the places into different combinations of dominant activities including those which pertain to agriculture production and land use. This method ultimately minimises the changes of over-simplified generalizations.

The present agriculture pattern is of intensive subsistence family farming type with an emphasis on food crops. Although Himachal Pradesh does not occupy any important place in food production among the states of India, it illustrates the achievements of subsistence farmers in modifying agriculture to raise their socio-economic conditions. The twin factors of hilly terrain and great diversity of agro-climatic conditions exercise a great control on agriculture. The result in small size holdings, difference in crop-timings with a change in altitude and variety in methods and types of farmings. Being the major existing resource, agriculture needs a thorough study within a regional frame, laying down the basis of cropping region. For this purpose the crop-combination regions have been carried out in Himachal Pradesh. The importance of such analysis as based on:[13]

(a) That crop-combination is comprehensive and provides better understanding of agriculture mosaic of an agro-climatic region.

(b) It shows the trend towards the specialization, e.g. increased specialization of horticulture is due to poor performance of food crops. Competitive areas with poor quality crops, new horticulture technology has transformed the temperate belt, turning it from mixed farming to the specialized temperate fruits mainly apples.
(c) It helps to understand the agriculture restructuring. Farming has became a more attractive financial institution which can combine profits from farming with capital gains from appreciating value of land.
(d) It highlights the accumulation process and regional differences. This is mainly happening in horticulture development in Himachal Pradesh.
(e) It further helps to understand the process to accumulation changes the traditional relationship, economic growth and transformation.
(f) It provides insight to relative profitability of the production process due to transformation.
(g) It shows the dependency of the farmer community on agriculture.
(h) Crop-combination has to highlight the relationship with industrial development mainly in agro-based industries.
(i) It also has become the important tool in planning.

The crops are generally grown in combinations and it is rarely that a particular crop occupies a position of total isolation from other crops in a given real unit.

With these inherent advantage, in studying crop-combination people from various disciplines have been attracted to study these association in a well defined manner.

Different methods applied in the delineation of crop combination region can be summed up under two categories—arbitrary choice method and statistical based method. The statistical techniques about crop combination have been modified by the geographers from time to time, viz. Doi (1959), Thomas (1963), Coppock (1964), Dikshit (1973) and Singh (1977).

For the present study Rafiullah (1965), 'New Maximum Positive Method' devised in "A New Approach to the Functional Classification of Towns" has been used. It is developed on Weaver's Method, which had some inherent weakness of "Crops in the series by the resultant combinations became over generalized". The technique devised by Rafiullah may be expressed algebraically as follows:

$$d = \sqrt{\frac{\hat{A}\ D_p^2 - D_n^2}{N^2}}$$

Where 'd' is the deviation, $D_p$ is the positive difference and $D_n$ is the negative difference from the medial value of the theoretical curve value of the combination, and N is the number of functions or Crops in the combination. For the present analysis 22 food and non-food crops have been included to see the crop-combination in Himachal Pradesh. They are: Wheat (W), Rice (R), Maize (M), Barley (B), Ragi (Rg), I. Millet (I.M), Common and Little Millet (CL), Varagu (Vr), Gram (G), Green Gram (GG), Black Gram (BG), Red Gram (RG), Horse Gram (HG), Masur (Ms), Other Pulses (OP), Apple (Ap), Mangoes (MG), Potatoes (Pot), Onion (On), Fresh Fruits (FF), Chillies (C), and Ginger (Gg).

The main purpose to include so many crops was that Himachal Pradesh has shown high rate of transformation in agriculture. Farmers are diverting from food crops to horticulture and vegetable farming on commercial basis. At this stage area under such fruits and vegetables may not be very high but it provides higher income value to the farmers. In the earlier chapter we have seen that areas under fruits and vegetables have shown high positive rate of change from 1970-71 to 1988-89. These crops have made significant impact on the local scale in transforming the socio-economic status of the farmers.

## Crop-Combination Regions in 1970-71

On the basis of a real strength of individual crops, the crop combination regions are formed by using Rafiullah's (1965) Maximum Positive Deviation Method. A large number of crops have come in combination but in total five combination have been identified which, in fact, form 12 contiguous units.

(1) *Mono-Culture*: Mono-culture as seen in the form of one crop region is not widely spread in Himachal Pradesh, but it is only confined to one tehsil—Spiti in the temperate dry region. The crop Barley does exist a much powerful influence in the temperate wet and temperate dry region of the State. The physio-climatic diversity

and edaphic factors are chiefly responsible for making this are responsible for growing Barley.

(2) *Two Crop-Combination*: There are two crops—Wheat and Maize—which make combination. There are three compact clusters, the first one is in the Shiwalik hills, second in sub-temperate (mid-hill) and third one is in temperate dry (higher hills). The first one extends in the sub-tropical region covering 13 tehsils out of 15 tehsils as well as in sub-temperate region it occupies 4 tehsils. The second region is in the North-western part of the State lies in sub-temperate region covering only one tehsil of Chamba district. The third cluster is local in temperate dry region in Kinnaur district covering 5 tehsils.

(3) *Three Crop-Combination*: There are 8 crops, Wheat, Maize, Rice, Barley, Common Millets, Potatoes, and Apple, which make combination with one or the other. Such areas where 3 crops dominate together quite compact. The major region is in the central part of the State comprises 20 tehsils. It covers most of the area in sub-temperate region which alone covers 11 tehsils out of 15 tehsils. The crops in combination are Wheat, Maize, Rice, Barley, Potato. In the temperate wet region 6 tehsils constitute 3 crops combination with Wheat, Maize, Barley, Apple and Common Millets. Whereas in temperate dry region 3 crop-combination of Common Millets, Barley, Wheat, existing in 3 tehsils. The second of third small region constitute 3 crop-combination of Wheat, Maize, gram is located in sub-tropical region.

(4) *Four Crop-Combination*: There are 6 crops, Wheat, Maize, Barley, Common Millets, Potatoes and Apple, which makes combination with one or the other. Such areas where 4 crops dominate together is quite scattered. It forms 4 small zones, out of which 3 fall in temperate wet region and the fourth one is located in temperate dry region out of 51 tehsils in the State only 5 of them have shown 4 crop-combination.

(5) *Six Crop-Combination*: There are 8 crops, Wheat, Maize, Barley, Rice, Common Millets, Potatoes, Apple and Ragi, which makes combination with one or the other. There is only one small region in the South-central part of the State comprising 2 tehsils of Shimla district which lies in temperate wet region (see Table 19).

**Crop-Combination in 1988-89**

Over the period of time after the formation of new State in 1970-71 there has been planned development of agriculture and horticulture in the state. There has been lot of emphasis on development of various crops and fruits in different agro-climatic zones. Hence, in 1988-89 new crop combinations have emerged in the state. There is a total absence of mono-culture region which clearly indicate the heterogeneity and complexity of crops in the state.

*Two Crop-Combination*: There are 6 crops, Wheat, Maize, Apple, Potatoes, Barley and Common Millets, which make combination with one or the other. Such areas where two crops derivative together are quite compact forming two bigger regions and 3 smaller regions. Out of 51 tehsils in the state 28 tehsils comprise 2 crop combination. In the sub-tropical region food crops make combination but in sub-temperate and temperate dry region food crops have started making combination with horticulture crops. Out of 15 tehsils in sub-tropical region 13 tehsils make 2 crop combination making a compact zone running from North-west to South-east. In the sub-temperate region the South-eastern part has 4 tehsils one in the central part and one in the North-west. There are 4 tehsils in temperate wet region. In the temperate dry region there is one compact zone comprising of 4 tehsils running North-west to East direction. In the region the horticulture crops have started dominating with potatoes and apples.

*Three Crop-Combination*: There are 6 crops, Wheat, Rice, Maize, Barley, Apple and Common Millets, which make combination with one or the other. There are 15 tehsils farming 3 crop-combination. Such areas where 3 crop-combination exists form one bigger compact zone of 13 tehsils. Another two small regions of one tehsil each are scattered in the state. The biggest region is in the Central part of the State extended into 3 zones. There are two tehsils in sub-tropical zone followed by 7 tehsils in sub-temperate, 4 tehsils in temperate wet and one tehsil in temperate dry region in the state. In sub-tropical and sub-temperate region food crops are making 3 crop combination whereas in temperate wet and temperate dry region food crops are making combination with horticulture crops.

*Four Crop-Combination*: There are 6 crops, Wheat, Apple, Maize, Common and Little Millets, Potatoes and Varagu, which make combination with one or the other. Such areas of 4 crop-combination are quite scattered forming 5 small zones. Most of them comprise one tehsil and only one region has 2 tehsils. There is one tehsil in sub-temperate zone, 2 tehsils in temperate wet and 3 tehsils in temperate dry region. The comprising combinations are both agriculture and horticulture basis.

*Five Crop-Combination*: There are 6 crops, Wheat, Apple, Maize, Common Millets, Barley and Pulses, which make combination with one or the other. Such areas where 5 crops make combination are quite scattered. There are two small region comprising of one tehsil each are in the South-eastern part of the State. One region lies in temperate wet and the other fall in temperate dry region of the state. New crop in the form of pulses have shifted making foothold in the dry region of the State (see Table 15).

**Crop Combination—A Regional Analysis**

Before becoming a separate state in 1971 the state of agriculture was in a bad shape. It was all because of local needs the crops were grown in the State. But after the formation of state many changes have taken place.

The physio-climatic condition continues to influence the agriculture in Himachal Pradesh. The crop combination have emerged in a well defined manner according to the physio-climatic regions.

(1) *Sub-Tropical Region*: Out of 14 tehsils, Bhattiyat, Una and Nalagrah have three crops in combination rest all contribute 2 crops combination. Wheat and Maize are the dominant crop-combination. In Nalagrah and Una gram has combined with Maize and Wheat combination. In Bhattiyat Rice made a combination with Maize and Wheat.

*Sub-Temperate Region*: Wheat and Maize are the dominant crop combination in 5 tehsils of this region. Rice in the Valley and low lying area forms a combination with Wheat. In about 8 tehsils such situation has occurred. Although, Barley is not very preferred crop but still in Chachhyot Barley is grown with Wheat and Maize. In Theog vegetables has made a significant contribution.

*Temperate Wet Region*: In this region variety of crops combination occur. Out of 15 tehsils 7 tehsils make 3 crop combination of Wheat, Maize, Barley. But in Katkaif and Kumarsain apple cultivation have made a combination with Wheat and Maize. Another 4 tehsils, viz. Bharmaur, Kullu, Chopal and Theog have 4 crop-combination. The highest 6 crop-combination is found in Rohni and Chopal tehsils of Shimla district. They have a combination of fine and coarse grain crops with fruits and vegetables. Only one tehsil Churah has 2 crops combination.

*Temperate Dry Region*: This region by location is not very suitable for agriculture. Barley and Common and Little Millet is mainly grown in this region. It form 1 mono-culture region in Spiti tehsil. In all tehsils of Kinnaur district Common Millets occupy the largest area. Meanwhile, Potatoes have started making combination with these crops. It is the potential area for dry fruits and off-season vegetables.

**Crop-Combination—A Regional Analysis, 1988-89**

After the planned development of agriculture and horticulture new combination have emerged in 1988-89.

In sub-tropical region Nurpur and Bhattiyat tehsils have shown increase. Whereas, Nalagrah and Una have decreased from 3 crops-combination to 2 crops-combination. Earlier gram was grown in this area, now in 1988-89 it has shifted down—does not become a part of combination. This is heading towards specialization in Wheat and Maize crops.

In sub-temperate the change is marked in Jogindernagar and Theog tehsils where the combination has shifted from 2 crops at over 1970-71 situation. Rest the crop-combination has remained same with minor adjustment between Barley and Rice crops.

The most dynamic region is shift, showing crop-combination in the temperate wet region. Out of 14 tehsils, 10 tehsils have shown changing in cropping. The area is becoming more specialized towards raising horticulture crops. The food crop have started making combination with apple and potatoes. Even, in the lower hierarchy the variety of fresh vegetables have made a significant change.

The temperate dry region is the outmost zone in Himachal Pradesh, which is least developed in socio-economic condition and in agriculture. But due to its physio-climatic condition it has lot of potential for growing apple and off-season vegetables and potatoes. Over 1970-71 and 1988-89, 8 out of 9 tehsils have shown shift in crop combination. Barley and Common Millets are the only food crops grown in this region. Rest most of the area is devoted to horticulture.

**Dynamic Area of Himachal Pradesh**

After analyzing the shifting trend in crop-combination, temperate wet region and temperate dry region emerge as the dynamic region. Both these regions are heading towards specialization of fruits mainly apple and off-season vegetables with potatoes. Out of 23 tehsils in these two regions 18 tehsils have shown shifting trend in the crop combination whereas on the other hand in sub-tropical and sub-humid regions 5 tehsils have marked shift out of 28 tehsils. One thing is highlighted that is fine crops, e.g. Wheat have increased their share over Barley and Maize.

**Table 19:** Transformation in Crop-combination over 1970-71 in 1988-89.

| *Tehsil* | *1970-71 Crop-comb.* | *Name of the crop* | *1988-89 crop-comb.* | *Name of the crop* | *Shift* | *Remarks* |
|---|---|---|---|---|---|---|
| Nahan | 2 | Wheat, Maize | 2 | Wheat, Maize | No shift | |
| Ponta | 2 | Wheat, Maize | 2 | Wheat, Maize | No shift | |
| Rainuka | 2 | Wheat, Maize | 2 | Wheat, Maize | No shift | |
| Pachhad | 2 | Wheat, Maize | 2 | Wheat, Maize | No shift | |
| Sadar | 3 | Wheat, Maize, Rice | | Wheat, Maize, Ragi | No shift | Rice crop replaced by Ragi |
| J. Nagar | 3 | Wheat, Maize, Rice | 2 | Wheat, Maize | 3-2 | Share of Rice crop has gone down |
| S. Ghat | 3 | What, Rice, Maize | 3 | Wheat, Maize, Rice | No change | Maize replaced Rice in second position in 1988-89 |
| Chachiot | 3 | Wheat, Maize, Barley | 3 | Maize, Wheat, Barley | No change | Mize replaced Wheat in first position |
| S. Nagar | 3 | Wheat, Maize, Rice | 3 | Maize, Rice, Wheat | No change | Maize shiftes to 1 Rice to 2 and Wheat which was at 1 has gone to third position |
| Karsog | 3 | Wheat, Maize, Rice | 3 | Wheat, Maize, Apple | No change | Fruis have come in combination with food crops, Rice not very attractive |
| Rampur | 3 | Wheat, C. Maize, Barley | 4 | Wheat, Apple, Maize, C. Millets | 3-4 | CL 2-4 Barley-remained at the same position Apple moved to 2 position. |
| Rohru | 6 | C. Millets, Ragi, Wheat, Barley, Maize, Rice | 5 | Wheat C. Millets, Apple, Maize, Barley | 6-5 | Specialisation —Wheat 3-1 Ragi—Not grown Maize 5- C.M. 1-2 Apples –3 |

(*Contd.*)

*(Contd.)*

| *Tehsil* | *1970-71 Crop-comb.* | *Name of the crop* | *1988-89 crop-comb.* | *Name of the crop* | *Shift* | *Remarks* |
|---|---|---|---|---|---|---|
| Inbbal | 6 | Wheat, Pot, Rice, C.L. Maize, Ragi | 2 | Wheat and Apple | 6-2 | Very high specialisation Wheat with Apples |
| Chopal | 4 | Wheat, Maize, C. Millets, Ragi | 3 | Wheat, Maize, Apple | 4-3 | Wheat, Maize-same, Apple-3 fruit specialisation |
| Theog | 3 | Wheat, Maize, Potatoes | 4 | Maize, Wheat, Apple, Potato | 3-4 | Shift in Maize and Wheat, Apple-3 Pot-4 |
| Shimla | 2 | Maize, Wheat | 2 | Maize, Wheat | No change | Apple and Potatoes have moved to higher order |
| Suni | 2 | Maize, Wheat | 2 | Maize, Wheat | No change | |
| | 3 | Wheat, Apple, Maize | 3 | Apple, Wheat, Maize | Shift | Apple 2-1 Maize-same, Wheat 1-2 |
| K. Sain | 4 | Wheat, Apple, Maize, Potato | 2 | Apple, Potato | 4-2 | Highly commercialised no food crops in the combination |
| Kotkhai | 4 | Wheat, Maize, Barley, Rice | 3 | Wheat, Maize, Apple | 4-3 | Specialisation fruits—Apple in combination |
| Kulla | 3 | Wheat, Maize, Barley | 3 | Wheat, Maize, Barley | No change | |
| Baijar | 3 | Wheat, Barley, Maize | 2 | Wheat, Maize | 3-2 | Barley not in combination |
| Anni | 3 | Wheat, Maize, Barley | 4 | Wheat, Maize, C. Millets, Varagu | 3-4 | C. Millets and Varagu has taken 3 & 4 position |
| Mchar | 3 | C. Millets, Barley, Wheat | 2 | C. Millets, Barley | 3-2 | Non-remunerative crops specialisation |
| Sangla | 2 | C. Millets, Barley | 2 | C. Millets, Apple | No change | Fruit cultivation, specialisation |
| Kalpa | 3 | C. Millets, Barley, Wheat | 4 | Apple, Barley, C. Millets, Wheat | 3 –4 | Horticulture specialisation. Not very paying unit C. Millets. 1-3 Wheat 3-4 |

(*Contd.*)

*(Contd.)*

| *Tehsil* | *1970-71 Crop-comb.* | *Name of the crop* | *1988-89 crop-comb.* | *Name of the crop* | *Shift* | *Remarks* |
|---|---|---|---|---|---|---|
| Moorang | 2 | C. Millets, Barley | 5 | Barley, C.L. Apple, Opul, Maize | 2-5 | Diversification to five corps and horticulture |
| Poo | 2 | C. Millets, Barley | 3 | Barley, Apple, C.L. | 2-3 | Diversification to fine crops and horticulature |
| Hanrang | 4 | Barley, C. Millets, Wheat, Potato | 4 | Barley, Wheat C. Millets, Potato | No change | Internal shift between Barley and Wheat |
| Dehra | 2 | Wheat, Maize | 2 | Wheat, Maize | No change | |
| Kangra | 3 | Wheat, Maize | 3 | Wheat, Rice, Maize | No change | |
| Nurpur | 2 | Wheat, Rice, Maize | 3 | Wheat, Maize, Rice | 2-3 | Rice is becoming important crop in valley regions |
| Palampur | 3 | Wheat, Maize | 3 | Wheat, Rice, Maize | No change | |
| Chamba | 3 | Wheat, Rice, Maize | 2 | Maize, Wheat | 3-2 | Barley is not preferred being a non-renumera-tive crop |
| Chuarah | 2 | Maize, Wheat, Barley | 2 | Maize, Wheat | No change | |
| Bhattiyat | 3 | Maize, Wheat | 3 | Maize, Wheat, Rice | No change | |
| Bharmaur | 4 | Maize, Wheat, Rice | 3 | Wheat, Maize, Barley | 4-3 | C. Millet is not grown |
| Pangi | 3 | Wheat, C. Millet, Maize, Barley | 4 | C.L. Wheat, Barley, Maize | 3-4 | Wheat 3-2, Barley 2-3, Meagre out |
| Sadar | 2 | C. Millets, Barley, Wheat | 2 | Maize, Wheat | No change | |
| Ghum-armin | 2 | Maize, Wheat | 2 | Maize, Wheat | No change | |
| Nalagrah | 3 | Maize, Wheat, Gram | 2 | Maize, Wheat | 3-2 | Gram (pulses not grown in commination) Wheat replaced Maize. |
| Kashli | 2 | Wheat, Maize | 2 | Maize, Wheat | No change | Order has changed |
| Solan | 2 | Maize, Wheat | 2 | Maize, Wheat | No change | |

(*Contd.*)

*(Contd.)*

| *Tehsil* | *1970-71 Crop-comb.* | *Name of the crop* | *1988-89 crop-comb.* | *Name of the crop* | *Shift* | *Remarks* |
|---|---|---|---|---|---|---|
| K. Ghat | 2 | Maize, Wheat | 2 | Maize, Wheat | No change | |
| Arki | 2 | Wheat, Maize | 2 | Maize, Wheat | No change | |
| Hamirpur | 2 | Wheat, Maize | 2 | Wheat, Maize | No change | |
| Badsar | 2 | Wheat, Maize | 2 | Wheat, Maize | No change | |
| Una | 3 | Maize, Wheat, Gram | 2 | Wheat, Maize | 3-2 | Pulses not in combination |
| Amb | 3 | Maize, Wheat, Gram | 2 | Maize, Wheat | 3-2 | |
| Lahaul | 3 | Barley, Wheat, C. Millets | 2 | Potato, C. Millets. | 3-2 | Vegetable specialisation |
| Spiti | 1 | Barley | 2 | Barley, Potato | 1-2 | Vegetable specialisation |

*Source*: Complied by the Researcher.

## Findings

The main findings on the basis of trend of change in area are as follows:

- The total area under food and non-food crops has shown positive trend of change in the state, especially in sub-tropical, sub-humid and temperate wet region. The negative transformation is recorded in temperate-dry region.
- The agricultural transformation during kharif and rabi seasons is marked by positive trend of change in the state but due to physio-climatic conditions and techno-institutional changes, the agriculture is highly variable in different zones.
- The positive rate of change under food crops in kharif season is recorded in sub-tropical, sub-humid and temperate-wet regions. The temperate-dry region depicted negative rate of change. The non-food crops registered negative rate of change in the state during this season. This pattern is found in sub-humid and temperate-wet region. The sub-tropical region did not register any change over the period of time.
- In the rabi season food crops registered negative transformation in the state. The sub-humid region highlighted positive change and temperate-wet and dry region showed

negative rate of change. At the same time, non-food crops during this period recorded a high positive rate of transformation in the state as well as in the region.

- The positive transformation in the state and in regions is observed in the area under cereal cultivation.
- The same pattern of cereal transformation is not found when individual crops are taken into consideration. The area under rice cultivation has declined in the whole state as well as in all regions. The area strength of wheat at the state level and in sub-humid, temperate, wet and dry region has shown declining trend, while only one sub-tropical region has recorded positive change. The area under barley, ragi and common and little millets has also followed the same trend as registered in wheat at the state and regional levels. At the same time, area under maize cultivation has increased in the state as well as in the regions.
- The area under pulses has shown high declining trends in the state as well as in all the regions. The same trend in followed by black gram but area under measure has increased at the state and regional levels.
- Wherever the area under cereal cultivation has increased, the area under pulses has declined at the state and regional levels.
- Horticulture is becoming important and as a result total area under fruits and vegetables has shown very high positive transformation at the state and tehsil level.
- The areas showing negative transformation in total foodgrain have registered high positive transformation in area under fruits and vegetables.
- The crop-combinations have emerged in a well-defined manner according to the physio-climatic conditions. Now horticulture crops are making combination with food crops, which never existed earlier.
- Two-crop combination of maize and wheat dominate in the sub-tropical region while at places rice and pulses are seen making combinations.
- The sub-humid region constitutes rice-wheat combination in low-lying locations while wheat and maize in upland areas. At places the vegetable are seen as making combinations with food crops.

- The temperate-wet region has 3 corp-combination of wheat, maize and barley. But now new combinations have emerged with apple and vegetables.
- The temperate dry-region has not shown much transformation and, hence, this region continues to form 2 crop-combination of barley and common and little millets.
- The sub-tropical and sub-humid regions specialise in fine crops, whereas and temperate wet and dry region continue to dominate in coarse crops.
- As a result where there had been negative transformation in area under foodgrains the crop-combinations has both changed. The positive transformation in area under food-grains did not show any definite pattern in crop-combination.
- The temperate-wet and dry region constituting 23 tehsils out of which 18 have marked change in crop-combination and in sub-tropical and sub-humid region 23 tehsils out of 28 tehsils have registered no change. Hence, the farmers too are very dynamic as compared to the later two regions.

## NOTES

1. Schultz, T. W. (1964): *Transforming Traditional Agriculture*, Ludhiana, India.
2. Weitz, R. (1971): *From Peasant to Farmer: A Revolutionary Strategy for Development*, New York, pp. 6-9.
3. Todaro, M.P. (1993): *Economic Development in Third World*, Orient Longman, Hyderabad, India, pp. 300.
4. Arnon, I. (1987): *Modernisation of Agriculture in Developing Countries—Resource Potential and Problem*, U.K.
5. Singh, B.B. (ed.), (1992): *Transformation of Agriculture and Business Enterprises*, Ghaziabad, India.
6. It will form a linear treand on the simple equation; YT = a + bx. In this equation YT increases by constant absolute amount per annum. This amount is given by efficient 'b'. Many other types of trends are available. A commonly used one is where YT increased not by a constant absolute amount per annum, but by a constant percentage per annum. Such trend is defined by YT = ABX.

   Where A is constant and B is exponential co-efficient. For example if B = 1.01, then YT increases by 1 per cent per annum. This type of trend is called exponential trend.
7. Annual Season and Crop Report, 1985-86, *Directorate of Land Receipts*, Himachal Pradesh, Shimla.

8. Singh, S. (1981): *Dynamics of Cropping Pattern in Northern India in Perspection in the Nature of Agriculture Land use Planning.* Geog-Noor Mohammed (ed.), pp. 485.
9. Unpublished Mphil. Thesis, K. K. Shonu.
10. Annual Season and Crop Report for the year 1985-86, *Directorate of Land Records*, HP, Shimla.
11. Joshi, K.L., (1984): *Geography of Himachal Pradesh,* National Book Trust of India, New Delhi.
12. Weaver, J.C., (1954): "Crop Combination Regions in Middle West", *Geographical Review*, Vol. XLIV, No. 2, p. 176.
13. L.L. Pownwell (1953), Standard Deviation Method—H.J. Nelson (1955), Mean Deviation Method—J.C. Weaver (1954), Ayyar (1969), H. Singh (1965), M. Husain (1972), Least Square Deviation Method—Thomas (1963), Maximum Positive Deviation Method—S.M. Rafiullah (1965), Modified Minimum Deviation—Kikukazi Doi (1959), M.F. Siddhqui (1967), N. Mohammed (1975), A. Majeed (1978), A.H. Khan (1973), Maximum Distance Method—N.P. Ayyar (1969), Lower Unit Method—A.G. Athqwale (1966) and Successive Quotients Method—Koztowicki (1972).

## REFERENCES

Athawale, A.G., (1966): "Some New Methods of Crop Combination", *Geographical Review of India*, Vol. XXVIII, No, pp. 29-33.

Ayyar, N.P., (1969): "Crop Regions of Madhya Pradesh: A Study in Methodology", *Geographical Review of India*, Vol. XXX, No. 1, p. 17.

Chisholm, M., (1966): *Geography and Economics*, Hutchinson & Co., London.

Doi, K., 1959: "The Industrial Structure of Japanese prefectures", *Proceeding of International Geographical Union, Regional Conference in Japan*, 1957, pp. 310-316.

Kostrowicki, J., (1972): "Some Methods and Techniques to Determine Crop and Other Land use Combination as used in the polish land use studies", *in Symposium on Land use in Developing Countries, 21St International Geographical Congress, 1968* (ed, M. Shafi, M. Anas and M.F. Siddiqui) Department of Geography, A.M.U., Aligrah, pp. 83-88.

Mohammad, N., (1975): "Crop Association Region in Ghaghara-Rapti Doab", *the North-eastern Geographer,* Vol. VII, No. 1 and 2, pp. 3-9.

Molnar, J. J. *et al.*, (1986): *Agriculture Change—Cosequences for South Farmers and Rural Communities*, West-View, USA.

Munton, R.J. C., (1969): "The Economic Geography of Agriclutre", In Cooke and James H. Johnson (ed), *Trends in Geography*, Pergamon, London, pp. 143-182.

Nelson, H. J., (1955): "A Service Classification of American Cities", *Economic Geography,* Vol XXI, No. 3, pp. 189-210.

Pownall, L.L. (1953): "The Function of New Zealand towns", *Annals of the Association of Americn geographers,* Vol. XLIII. pp. 332-350.

Rafiullah, S. M., (1965): New Approach to Functional Classification of Towns", *The Geographer,* Vol. XII, pp. 40-53.

Reeds, L.G., (1964): Agricultural Geography: Progress and Prospects", *Canadian Geographer,* Vol. 8, pp. 52.

Rene, L., (1992): *Modern Plantation Agriculture-Corporate Wealth and Labour Squator,* Zed Books Ltd., London.

Schultz, T. W., (1964): *Transforming Traditional Agriculture,* Lyall Book Depot, Ludhiana.

Sharma, K.K., (1987): *Development of Horiculture in Shimla District,* Unpublished M. Phil Dissertation submitted to Department of Geography, University of Delhi.

Singh, G.B., (1979): *Transformation of Agriculture—A Case Study of Punjab,* Vishal Puplication, Kurukshetra.

Tiwari, P.D. and Jain, C.K., (1989): *Modernisation of Agriculture and Food Arallability in India,* Northern Book Centre, New Delhi.

Thomas, D. (1963): Agriculture in Wales During Napoleonic Wars", (cardiff), pp. 80-81, cf. Coppock, J.T. (1964): *The Agricultural Atlas of England and Wales,* (London), pp. 211-212.

Vashistha, S.B., (1989): *Farmers training for Agricultural Development in India,* Deep and Deep, New Delhi.

Weaver, J.C. (1954): "Crop-Combination Regions in Middle West", *Geographical Review*, Vol. XLIV, No. 2, p. 176.

# 8

# Human Resource and Agricultural Development of Lumbini Zone, Nepal

S.B. SINGH, B.R. KARERIYA and SHYAM B. SINGH

## Introduction

The nature endowment comes under resource category when the man thought his mental and physical ability gets benefited through proper use of available materials in that region. The physical factor provides platform upon which development process proceeds ahead in the condition with technological advancement of man. From this, it is obvious that development hinges much on the response of people. Efficient people may make best out of given resources, while inefficient people may waste them. Hence economic development under given resources is determined largely by human behaviour, human quality and human institutions (Shrestha, 1998). Joseph Fisher (1962) has rightly remarked that natural resources are in themselves passive in their influence on economic growth. They remain dormant and largely unknown till such time as the general including technology recalls their use value. Then acquire a dynamic character largely through changes introduced by man. Further, it is related to sustainable development of both natural as well as human resources. So the concept of sustainable development, involves rational and successful management of resources, either natural or human through technology and efficient energy systems to satisfy human needs, by maintaining or enhancing the quality of environment. The production of wealth and their equitable distribution and protection of environment are equally relevant for improving the quality of human life and fragile ecosystem (Singh, 1995). There is growing awareness about the need to substitute current production system with a more sustainable and integrated system of agriculture production. The gravity of the environmental degradation question has lead the scientists, planners and geographers to examine ecologically sound, economically viable and socially sustainable farming system for different agro-climatic regions.

**Profile of the Study Area**

The study area, Lumbini Zone lies at the Southern part of the Western Development Region of the Kingdom of Nepal between 27°17′N to 28°17′N and 82°43′E to 84°23′E with 8,975 km$^2$ area and 20,13,673 population (Census 1991) and 25,26,868 (Census 2001). It consists of six districts, namely Arghakhanchi, Gulmi, Palpa, Nawalparasi, Rupandehi and Kapilvastu. Of which, former three districts fall in the hill region while later, three ones are in the tarai region. Thus, apparently it can be said that this zone is extended from hilly region, the north to Indian border, the south, she makes her zonal boundary from three sides, to the East Narayani, to the West Rapti, and to the North Dhaulagiri and to the north-east direction there lies Gandaki Zone while southern boundary of this zone is separated internationally with India (Fig. 1). Geographically, it is located at the Central part of Nepal which has a considerable bearing on the gēology of Nepal as a whole, since the numerous southward thrusts, geographically and chronologically originate here. North of Pokhara lies the dominating heights of Annapurna Himalyan mountain the base of which represents the oldest root-zone of uplift in Nepal Himalaya (Hagen, 1960). The Paraautochthonus zone of Pokhara or so-called Pokhara Zone is considered the lowest unit in Nepal (Gurung, 1965). But the area, Lumbini Zone is located to the South of Pokhara Zone extending from hilly region, Mahabharat range (3000 m) to terai region (150 m). Between these two unequal topographical parts there exists the Churia range (900 m) running just South of the Mahabharat mountains are analogous to, and contemporary of the Siwalik ranges of north India.

Climatically, this zone is much varied as altitude increases from south to north; the summer and winter temperature are also different in their nature. During summer, temperature ranges from 30° to 20°C and average rainfall is about 200 cm. Terai is normally hotter than hill as it receives *Loo* (Heat wave) while in winter temperature decreases from 15°C to 10°C from north to south and it becomes moderately cool. During this period, cold waves locally known as

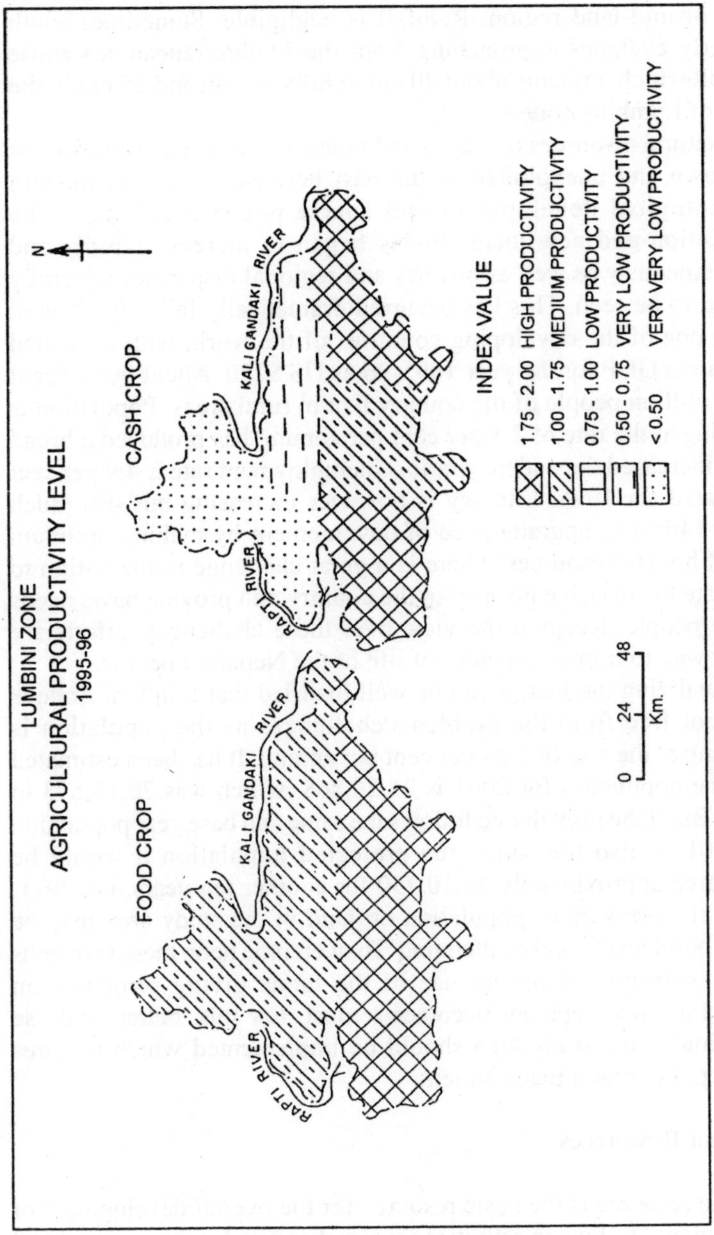


**Fig. 1:** Lumbini Zone: Agricultural Productivity Level

*Sirato* descend down from there and make weather very cold in most parts of mid-land region. Rainfall is negligible. Sometimes south westerly cyclones approaching from the Mediterranean sea cause rainfall which amounts about 40 cm in hilly region and 25 cm in the parts of Lumbini Zone.

Natural resources like flora and fauna, minerals etc. were almost unknown and unexploited in the past because of lack of modern technological development and sparse population, but as the population and new technologies began to increase rapidly and simultaneously as well as poverty and regional disparities generally started to be seen. This has occurred dramatically in Nepal. Nepal, being one of the developing countries of the world with estimated per capita GDP for the year 1995-96 as US $220. About 90 per cent of 21 million people of the country live in rural areas. Population is growing at the rate of 2.1 per cent per annum, has produced a broad base, increased dependency ratio. Unemployment rate is 4.9 per cent estimated for 1996. Literacy is however increasing but still much lower (40%) to upgrade pace of development by making optimum use of human resources. There is a great challenge to the nation to estimate the massive poverty in the country and provide basic needs to the people. Keeping the view over these challenges, efforts are under way to improve quality of life of the Nepalese people.

Realizing the fact, it can be well justified that Lumbini Zone is also not free from the problems/challenges as the population is growing at the rate of 2.84 per cent per annum. It has been estimated that the population for 2001 is 25,26,868, which was 20,13,673 in 1991. But if the growth rate follows the same and base year population of 2001 is also the same, the projected population it would be estimated approximately 35,10,750 for another ten years, i.e. 2011. Thus, the pressure of population on land in the study area may be much problematic and challenging. To overcome upon these problems and constraint the proper use of the lands giving priorities on agriculture development becomes a must. For this, better land use planning in the study area should be implemented which requires better policies and management.

## Human Resources

Human resource is the basic resource for the overall development of any region. The human activities are closely related to the geographical environment of the particular region. In other words, it can be stated that man is the essential resources as he utilizes resources and changes the natural environment. He is not only the beneficiary of the entire process of resource development and utilization (Zimmerman, 1951) but also is the most potent and dynamic agent of the production. This resource base is basically useful to mobilize and utilize the natural resources properly. Problems of unemployment and low productivity area, while literacy, marital status etc. are the factors that influence the quality of human resource. It is, therefore, needed precisely planned development policy that these problems should be well tackled and solved so that human resources would be changed into active force for growth and development (Singh, 1988). Human resource and agricultural development of any particular region are primarily based on the socio-economic facilities with their location at appropriate places serve the people and play an important role in the development of a particular geographical area (Singh and Singh, 1987).

**Population Distribution and Man-Land Ratio**

Man-land ratio is mainly affected by geographical factors such as terrain, climatic conditions, availability of water, availability of arable land and other socio-economic factors. The hilly districts of Lumbini Zone are sparsely populated which constitutes 2,08,391 population in Arghakhanchi, 2,96,654 population in Gulmi, 2,68,558 in Palpa, while the terai districts such as Nawalparasi possesses 5,62,870 population, Rupandehi and Kapilvastu possess 7,08,419 and 4,81,976 population respectively (Census, 2001). Terai districts, being feasible from economic point of view mainly with agricultural activities is economically viable. So, the concentration of the people here is higher as compared to the hilly region. Furthermore, this can be proved with the help of densities of existing districts of the region. Thus man-land ratio is the ratio of total population to the total area of the zone. In other words, it can be well justified from densities of that particular region. The 1981 Census data reveals that the people residing per km$^2$ in Arghakanchi, Gulmi, Palpa, Nawalparasi, Rupandehi and Kapilvastu were 131.8 persons/km$^2$, 207.2 persons/km$^2$, 156.18 persons/km$^2$, 142.8 persons/km$^2$, 278.2 persons/km$^2$ and 155.3 persons/km$^2$ respectively, while in 1991 the man-land ratio is higher as compared to 1981 and the above-mentioned districts possess 151.6 persons, 231.7 persons, 172.1 persons, 201.9 persons, 383.9 persons and 213.9 persons per km$^2$ respectively. Likewise, in 2001 the density is 174.6 persons/km$^2$ for Arghakhanchi, 258.1 persons/km$^2$ for Gulmi, 195.5 persons/km$^2$ for Palpa, 260.3 persons/km$^2$ for Nawalparasi and 520.8 persons/km$^2$ for Rupandehi and 277.3 persons/km$^2$ for Kapilvastu. The density of Lumbini Zone was 174.6 persons/km$^2$ in 1981 while it increased to 224.3 persons/km$^2$ in 1991 and 281.5 persons in 2001 (Table 1).

**Table 1:** Population Density in Lumbini Zone.

| *Area/District* | *Density (Persons/km$^2$)* | | |
|---|---|---|---|
| | *1981* | *1991* | *2001* |
| Arghakhanchi | 131.8 | 151.6 | 174.6 |
| Gulmi | 207.2 | 231.7 | 258.1 |
| Palpa | 156.18 | 172.1 | 195.5 |
| Nawalparasi | 142.8 | 201.7 | 260.3 |
| Rupandehi | 278.7 | 383.9 | 520.8 |
| Kapilvastu | 155.3 | 213.9 | 277.3 |
| Lumbini Zone | 174.6 | 224.0 | 281.5 |

*Source*: Population Census, 1981,1991 and 2001, Kathmandu.

Similarly, other types of densities such as physiographic, agricultural, caloric and nutrition have also been computed. Physiographic density is the ratio of total population to the total cultivated area, and densities 609.1 persons/km$^2$. Agricultural density is the ratio of total rural population to the total cultivated area and density is 568.7 persons/km$^2$ while caloric and nutrition densities are the ratio of total rural population and total population to the total food cropped area and densities are 500.3 persons/km$^2$ and 535.7 persons/km$^2$ respectively. (Table 2)

**Table 2:** Density Types in Lumbini Zone.

| *Types* | *Density (persons/km$^2$)* |
|---|---|
| Physiographic | 609.1 |
| Agricultural | 568.7 |
| Caloric | 500.3 |
| Nutrition | 535.7 |

*Source*: Population Census, 2001.

## Sex Ratio

Sex ratio is an essential element to analyse the human resources as it determinates the quantum of labour force supply and influence the socio-economic structure. In the study area of hilly districts: Arghakhanchi, Gulmi and Palpa the sex ratio is 1148, 1204 and 1141 respectively as per 1000 males while in the districts of terai such as Nawalparasi; Rupandehi and Kapilvastu, the sex ratio is 1003, 973 and 941 respectively as per 1000 males. But as a whole in Lumbini Zone there are 1036 females residing as per 1000 males numbers (Table 3).

**Table 3:** Sex Ratio in Lumbini Zone by District-wise.

| *District* | *Male* | *Female* | *Total* | *Sex Ratio* |
|---|---|---|---|---|
| Arghakhanchi | 84172 | 96712 | 180884 | 1148 |
| Gulmi | 121079 | 145536 | 266331 | 1204 |
| Palpa | 110325 | 125988 | 236313 | 1141 |
| Nawalparasi | 217731 | 218486 | 436217 | 1003 |
| Rupandehi | 264544 | 257606 | 522150 | 973 |
| Kapilvastu | 191444 | 180334 | 371778 | 941 |
| Lumbini Zone | 989011 | 1024662 | 2013673 | 1036 |

*Source*: Nepal Population, Census 1991.

Similarly, census 2001 reveals that the sex ratio in the hilly districts such as in Arghakhanchi is 1162, in Gulmi 1217 and in Palpa, it is 1147 as per 1000 miles. The terai districts' sex ratio is less than hilly districts. Nawalparasi has 1022 sex ratio per 1000 males, while, Rupandehi and Kapilvastu possess 963 and 944 sex ratio as per 1000 males. On the whole, it is justified that the sex ratio has been slightly increased in district-wise and decreased in Lumbini Zone, i.e. 1034 as per 1000 males as compared to 1991 census (Table 3A).

**Table 3A:** Sex Ratio in Lumbini Zone by District-wise.

| *District* | *Male* | *Female* | *Total* | *Sex Ratio* |
|---|---|---|---|---|
| Arghakhanchi | 96349 | 112042 | 208391 | 1162 |
| Gulmi | 133771 | 162883 | 296654 | 1217 |
| Palpa | 125068 | 143490 | 268558 | 1147 |
| Nawalparasi | 278257 | 284613 | 562870 | 1022 |
| Rupandehi | 360773 | 347646 | 708419 | 963 |
| Kapilvastu | 247875 | 234101 | 481976 | 944 |
| Lumbini Zone | 1242093 | 1284775 | 2526868 | 1034 |

*Source*: Nepal Population Census, 2001.

Thus, the sex ratio is higher in the hilly districts of the study area than the terai districts. It is because more male numbers remaining out of their homes for long periods for employment as well as job opportunities.

## Agricultural Intensity

Agriculture is the most important constituent of man's primary occupations and forms the first vital link of human interaction with his environment (Singh, 1995). Under the land use pattern agriculture is the lifeblood of any region. This is borne by fact that out of a total reported area of 8,97,500 ha., 4,66,230 ha. (51.95%) lands have been devoted to agriculture arable land. The whole study area covers the total geographical area, 8,97,500 ha. of which, 51.95 per cent is under agriculture. Besides, 3,54,171 ha. (39.46%) forest, 1806 ha. (2.01%) grazing lands/pastures, 42, F78 ha. (4.70%) roads and settlement and rest 16,855 ha. (1.88%) is not available for agricultural purposes.

It has been observed that all districts situated in the same ecological and agro-economic environment, record diversity in agriculture. The rational use of land at different levels show three types of intensities, i.e. cropping intensity (ratio of harvested area to arable area) land use intensity (ratio of gross sown area to net sown area) and irrigation intensity (ratio of gross irrigated area to net irrigated area) have been distinguished. The density of population is much greater and land resources available for extension of cultivation are much smaller. In the study area, cropping intensity varies from 105.29 (Rupandehi) to 149.14 per cent (Nawalparasi). But cropping intensity for whole Lumbini Zone is 119.16 while land use intensity varies from 102.77 in Rupandehi to 144.89 in Arghakhanchi and for whole Lumbini Zone is 112.54. Similarly, there is also variation in irrigation intensity as it is differentiated from 139.98 in Arghakhanchi to 249.60 in Kapilvastu. But for whole study area the irrigation intensity is 201.99 per cent (Table 4).

## Irrigation System

It has been well recognised that irrigation plays a vital role to farmers' prosperity and national agricultural production. The water supply is the need for a well-controlled year around without which it is virtually impossible to accelerate agricultural growth rates to a level of bringing

**Table 4:** Land use Character in Lumbini Zone.

| *Area/District* | *Cropping Intensity* | *Landuse Intensity* | *Irrigation Intensity* |
|---|---|---|---|
| Arghakhanchi | 125.10 | 144.89 | 139.98 |
| Gulmi | 142.65 | 108.05 | 190.74 |
| Palpa | 112.15 | 129.95 | 210.43 |
| Nawalparasi | 149.14 | 118.71 | 165.39 |
| Rupandehi | 105.29 | 102.77 | 195.14 |
| Kapilvastu | 108.19 | 103.39 | 249.60 |
| Lumbini Zone | 119.16 | 112.54 | 201.99 |

rural prosperity and economic transformation of the country. Although irrigation has been the centre piece of investment programmes in Nepal for more than forty years; the emphasis has been largely on expanding supplementary monsoon irrigation. Only 18 per cent of the arable area is commanded by well-controlled year-round water supplies (Nepal/APP,1995) because irrigation has to be dependent on the seasonal/monsoonal character of rainfall which generally occurs between 15th June to 15th October except for a few days during the winter season. The availability of rainfall is the main cause for the development of irrigation facilities in the area.

Thus, keeping the view, it can be well conceived that in the study area, the irrigation system is quite peculiar as terai region is drained by both groundwater system and river flowing from the north while hilly region is basically irrigated with terraced method and the perennial and seasonal rivers are the main source for these purposes from where streams/drains locally known as *Kulos* are made.

## Agricultural Productivity

The term production has been used to denote the per ha. yield of crops. It has been well understood that the terai districts of Lumbini Zone are more productive in both food crops as well as in cash crops as compared to hilly districts of the study area. It is because of the role played by some major factors, viz., plain area, accessibility, irrigation facilities, modernized implements used in farms, availability of transportation and marketing facilities and so on. In Lumbini Zone the overall production of food crops is 2.024.5 kg. per ha. while per capita production is 377.8 kg. Similarly the cash crop production per ha. in the study area is 13,799.6 kg. while per capita production is 234.9 kg. In the whole study area, Nawalparasi district has the highest cash crop production (18,053.1 kg./ha.) while 579.4 kg. is the per capita production and the lowest is possessed by Arghakhanchi district (2,229.7 kg./ha. and 18.2 kg./capita production). Some cash crops-such as oilseeds and sugarcane are also grown. Potatoes and other vegetables are grown mainly for household consumption. By substantially increasing agricultural and food production, these improvements would improve the food security situation in the region (Hirmiz and Karim, 1996). Hence to analyse the general agricultural productivity level in the study area at two points of time period, i.e. (1986-87 and 1995-96) production of respective crops has been taken into accounts which has further been depicted on the basis of the formula (Shafi, 1972):

$$\left[\frac{yw}{t}+\frac{yr}{t}+\frac{ymi}{t}-n\right]:\left[\frac{Yw}{T}+\frac{Yr}{T}+\frac{Ymi}{T}-n\right]$$

$$\frac{n\Sigma y}{t}:\frac{n\Sigma y}{T}$$

where

yw, yr etc. = total yield of the respective crops in the district.
Yw, Yr etc. = total yield of the respective crops at the national level.
t = area planted under the crop in the district, and
T = area planted under the crop at the national level.

The result exhibits that five types of agricultural productivity level which are high level of agricultural productivity (index value, 1.75-2.00). medium level of productivity (index value, 1.00-1.75) low level of productivity (index value, 0.75-1.00), very low level of productivity (index value, 0.50-0.75) and very low level of agricultural productivity (index value <0.50). Accordingly, the calculation justifies that during 1986-87, Nawalparasi having index value 1.40 and Rupandehi having index value 1.13 were under medium level of productivity while Gulmi, Palpa, Kapilvastu and Arghakhanchi possess 0.80, 0.80, 0.87 and 0.87 index value respectively and are put under the low level of productivity in reference to the food production. But this figure is varied from 1995-96. The all terai districts have fallen in medium level while the hilly districts are under low level of productivity followed by Gulmi (index value 0.70), Palpa (index value 0.84), Nawalparasi (index value 1.26) and Kapilvastu (index value 1.16) respectively (Table 5).

**Table 5:** Food Crop Productivity Level in Lumbini Zone.

| | | *1986-87* | | *1995-96* | |
|---|---|---|---|---|---|
| *Rank* | *Index Value* | *Name of District* | *Per cent* | *Name of District* | *Per cent* |
| I | 1.75-2.00 | | – | – | – |
| II | 1.00-1.75 | Nawalparasi, Rupandehi | 33.33 | Nawalparasi, Rupandehi, Kaplivastu | 50.00 |
| III | 0.75-1.00 | Gulmi, Palpa, Kapilvastu, Arghakhanchi | 66.67 | Gulmi, Palpa, Arghakhanchi | 50.00 |
| IV | 0.50-0.75 | – | – | | |
| V | <0.50 | | – | | |

*Source*: Statistical Year Book of Nepal, 1997.

Besides, in context of cash crops production in 1986-87, three districts—Nawalparasi (index value 2.41, Rupandehi (index value 1.79). Kapilvastu (index value 2.30) were in high level of productivity. Out of these, only one district, i.e. Gulmi, possessing the index value 0.98 was under the low level category while two districts Palpa and Arghakhanchi were categorised under very low level of productivity which possess 0.68 and 0.57 index value respectively. But this figure is completely varied as compared to 1995-96. During this period, two districts Nawal Parasi and Kapilvastu having 2.38 and1.79 index value respectively are under high level of productivity Rupandehi and Palpa have 1.66 and 0.58 index value are under medium level of productivity and very low level of productivity respectively. But two districts Gulmi and Arghakhanchi are categorized into very, very low level of agricultural productivity which possess <0.50 index value, i.e. 0.45 and 0.29 respectively (Table 6 and Fig. 1)

**Table 6:** Cash Crop Productivity Level in Lumbini Zone.

| | | *1986-87* | | *1995-96* | |
|---|---|---|---|---|---|
| *Rank* | *Index Value* | *Name of District* | *Per cent* | *Name of District* | *Per cent* |
| I | 1.75-2.00 | Nawalparasi, Rupandehi, Kapilvastu | 50.00; | Nawal Parasi, Kaplivastu | 33.33 |
| II | 1.00-1.75 | | – | Rupandehi | 16.67 |
| III | 0.75-1.00 | Gulmi, | 16.67 | | – |
| IV | 0.50-0.75 | Arghakhanchi, Palpa | 33.33 | Palpa | 16.67 |
| V | 0.50 | | – | Gulmi, Arghakhanchi | 33.33 |

*Source*: Statistical Year Book of Nepal, 1997.

## Epilogue

It is, therefore, concluded that there is an urgent need of proper human resource development providing them different types of opportunities for employment on the one hand, and suitable land use planning accentuating in agricultural development on the other hand. For this, the following steps should be taken into consideration:

1. Land use should be done in planned way but not haphazardly so that the interaction between rural and urban may be at the maximum level.
2. The empowering of agricultural development with special references to irrigation system, inputs and other innovative technologies should be done both in terai and hilly region of the study area.
3. Lumbini zone, as being an agricultural region, there should be another green revolution, white revolution (milk), blue revolution (fish) and yellow revolution (oil seeds), so that human resource and agricultural development may be possible further leading to enhancing the employment opportunities *vis-à-vis* poverty alleviation.

### REFERENCES

Fisher, J.L. (1962): *Role of Natural Resources in Economic Development: Principles and Problems*, London, p. 32.

Gurnng, Hark (1965): *Pokhara Valley, Nepal Himalaya: A Study in Regional Geography*, unpub. Ph.D. Thesis, University of Edinburgh, U.K., p. 17.

Hagen, Tony (1960): *Reports on Geological Survey of Nepal*, Zurich.

Hirmiz, Toma and A. Karim (1996): *Agriculture and Development in Western Asia*, Joint Publication of United Nations Economic and Social Food and Agriculture Organization of the United Nations, No. 18, Baghdad, p. 32.

HMG Nepal and Asian Development Bank (1995): *Nepal Agriculture Perspective Plan*, Inc. Washington, D.C., p. 69.

Shafi, M. (1972): Measurement of Agricultural Productivity of the Great Indian Plains, *The Geographer*. Vol. 19(1), pp. 4-3.

Shrestha, S.H. (1998): *Natural Resources, Economic Geography of Nepal*, Educational Enterprises, Pvt. Ltd., Kathmandu, p. 73.

Singh, B.N. (1988): *Integrated Rural Area Development and Planning*, Anupama Publication, Delhi.

Singh, S.B. (1995): Sustainable Development of Rural Environment: A Case Study of Suithakalan Block, U.P., *National Geographer*, Vol. 30(2), pp.123-130.

Singh, S.B. and Singh, B.N. (1987): Planning for Socio-Economic Facilities in Ghazipur Tehsil, U.P., *Nat. Geog. Jl. of India*, Vol. 33 (1), p. 89.

Zimmerman, W. (1951): *World Resource and Industries*, Harper and Row Publisher, New York, p. 91.

# 9

# Agricultural Land Use Debacles in Laos

JUPAKA MADHAVI

## Introduction

Lao People's Democratic Republic or Laos is a landlocked country, located in South-east Asia. It is located at 14° to 23° North latitude and 108° to 100° East longitude.[1] The local long form of Laos is 'Sathalanalat Paxathipatai Paxaxon Lao', with 18 administrative provinces[2] (1 special) and Vientiane as its capital. Laos derived independence from France on 19 July, 1949 and its constitution was promulgated on 14 August, 1991. In 1997 it became an Asean[3] member.

With a total land area of 236,800 sq. km. (including 6000 sq. km. of water) Laos runs a 5083 km. border with Thailand of 1,754 km. border in the West (with Mekong River forming a large part of its Western boundary), Vietnam of 2,130 km. border in West, Myanmar of 235 km. border in North-west, China of 423 km. border in the North and Cambodia having 541 km. border at its South. Most of the country is rugged and thickly forested having a tropical to sub-tropical monsoon climate with rainy season extending from May to November. Phou Bia (2,817 m) forms the highest elevation with the lowest point in South at 70 m where Mekong leaves the country. 1835 km of Mekong River on the Western border of the country flows from north to south providing 25 per cent of the fertile soil to Laos. Other rivers such as Tha, Ou, Ngum, Bangfai, Banghiang, and Kong provide immense potential for hydropower development and scope for irrigation in Laos.

With a total population of 6068117 people and a population growth rate at 2.44 per cent, Laos has the lowest population density in Asia of 26 people per square km[4]. Laos has up to 18,000 MW of hydroelectric power generating potential, of which just 620 MW has been developed to date.[5] Despite considerable fertile soil, low population density, abundant water resources, the land use pattern of Laos is deplorable with other predicaments like soil erosion, deforestation, floods, droughts and zero access to potable water.

## Agriculture

Laos has largely an agricultural economy. Around 50 per cent of its total GDP and approximately 80 per cent of the workforce is centred on the agriculture sector. Agricultural output growth has tended to be quite erratic from year to year, but has averaged 6.5 per cent per annum over the last decade[6]. 3.5 per cent of permanent crops to the total arable land. Rice is the single most (permanent) major wet season crop in Laos. Figure 1 depicts the rice production under different categories. As it is very evident from the diagram that only 9 per cent of rice is under-irrigated showing that the farmers are still dependent on rainfed rice production. It is in the Mekong valley itself that the country's main agricultural areas 'granaries' are found particularly surrounding Vientiane and Savannakhet area. Rice is cultivated even in uplands as more than 70 per cent of Laos is mountainous. Over the Bolaven Plateau much of coffee, tea and mountain rice is farmed.

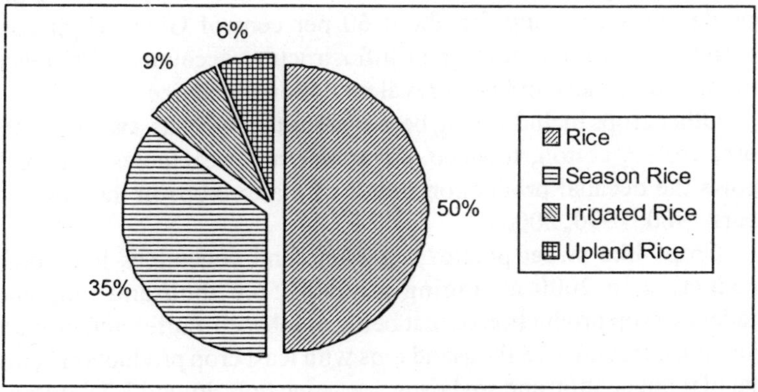


**Fig. 1:** Laos: Rice Production (in thousand tons), (2000).

Rice irrigation is rainfed, accounting for about 72 per cent of the cultivated area. Figure 2 depicts the harvested area of rice with respect to other crops in Laos. Around 2201.7 thousand tons of rice production in 719 thousand hectares of harvested area, with a yield of 3.1 tons per hectare has been estimated in 2000. Following rice is maize which is the second most important crop with its production being 117 thousand tons, and yield per hectare being 96.1 harvested in 49 thousand hectares.[7]

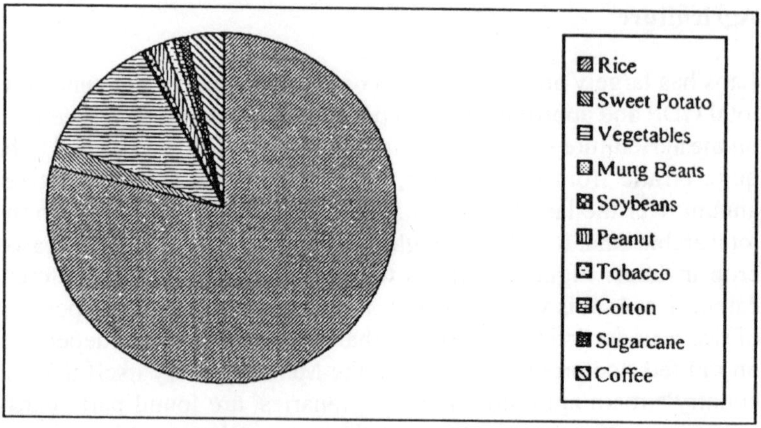


**Fig. 2:** Laos: Harvested Area of Major Crops (thousand tons), (2000).

In Laos more than 90 per cent of the rice farmers practice subsistence agriculture and they involve about 70 per cent of the population accounting for about 60 per cent of GDP.[8] High cost marketing and underdeveloped infrastructure accounts subsistence among the farmers and is a prevalent national practice.

Other crops include mung beans, peanuts, soybeans, sweet potato, corn, coffee, cotton, tobacco, sugarcane and vegetables. Figure 3 shows the decadal production trends of major crops in the selected years 1980, 1990, 2000.

Crops like sweet potato, sugarcane and vegetables have high production in 2000 averaging around 356.6 thousand tons and moderate crop production of that being of tobacco, coffee and peanuts averaging around 67.8 thousand tons with least crop production being that of mung, cotton and soybeans averaging only around 3.7 thousand tons. In 2000, except mung beans, all other crops have shown a

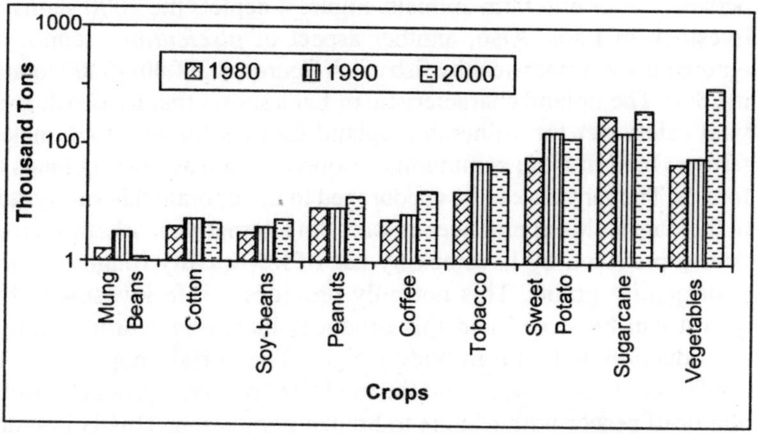


**Fig. 3:** Laos: Production of Major Crops (Selected Years 1980, 1990, 2000).

decadel growth from 1980 with intermittent growth fluctuations in 1990. However, since 1988, the shift to the market based economy has evolved by the farmers suggesting a proactive approach towards agriculture.

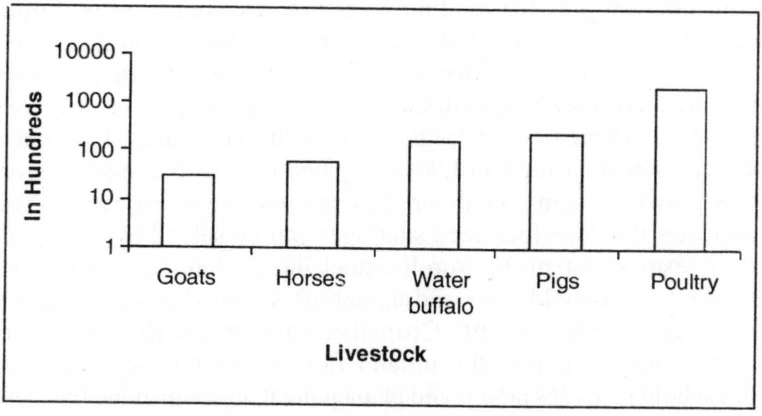


*Source*: CIA World Factbook 2004.

**Fig. 4:** Laos : Livestock (2004).

Livestock is a sub-system within the farming system of Laos just like any other farming system as there are many interactions between crops and farm animals. Figure 4 depicts the 2000 status of livestock in Laos. Also, another aspect of *pisciculture* cannot be ignored in Laos with healthy fish catch figures of 30000-40000 tonnes in 2000. The upland characteristic of Laos shows that livestock are a vital safety net for vulnerable upland farmers for livestock makes farmers benefit from community resources, such as grazing lands or forests. The farmers can be encouraged to incorporate this sub system in their agricultural practices as well. An alternative market for crops is also provided by livestock by use of low-quality roughages and poor-quality grains. This not only provides a safe investment but also it can be stored and this produces increased returns through reproduction and gain in body weight. The buffalo populations of 10000 show the supply and demand for livestock products: milk, and draff power with respect to human population. This is in much correspondence compared to pigs (1.3 million) and poultry (14 million) that dominate Laos's livestock with respect to supply demand parameters.

The impact of changes in production characteristics on the population of livestock shows the need of livestock and crop integration. For example, Cassava's (third most important crop in Laos, after rice and maize) roots are used mainly for human consumption and for feeding livestock, especially pigs. Young Cassava shoots are harvested for human consumption and recent livestock research has demonstrated that cassava can be a high potential livestock feed as dried cassava roots and leaves can be used as a feed supplement for pigs, cattle, fish and poultry. Cassava is widely grown throughout Laos by upland farmers but in small areas it is grown by using local varieties and with very few inputs. This indicates that livestock feed shortages can be solved by improving the livestock systems by crop-livestock integration. This can be done where the household, crop, and animal sub-systems become integrated and yet interdependent. Crop-livestock integration becomes mandatory to sustain the upland farming system such that the household provides labour and management, crops provide feed, and the animals generate power, manure, meat, milk, and capital while keeping in mind the upland or hinterland areas for they are remote, highly diverse, rapidly changing, heterogeneous with rampant degradation of natural resources.

However, livestock debacles such as pressure for farm labour with food crops, for land and high maintenance cost due to small number of animals on farms cannot be ignored. Not to forget, that often it's the livestock that is more prevalent to UXO accidents while grazing, which not only creates financial loss but causes out of cause pressure on the farm.

## Agricultural Debacles

There are several constraints to sustainable agriculture in Lao (PDR) of which most alarming are soil erosion, resource degradation, and on increasing population of the "poorest of the poor". Following are the agricultural exigencies in Laos:

1. *Irrigational Dependence on Monsoons*: The greatest constraint to rice production is the dependence on a rainfed rice-based system. More than 98 per cent of the area is rainfed and drought in rainfed areas amputatedly interrupts the agricultural process and agricultural growth. There has been very little adoption of improved technologies for irrigation in the rice-based ecosystems for only 9 per cent or the rice area is irrigated. An increase in the irrigated area could be achieved through the development of small irrigation systems. The government has targeted these areas for increased production to achieve agricultural self-sufficiency in Laos.
2. *Inferior Rice Quality*: Traditional varieties or rice dominate in more than 85 per cent of the area of rainfed lowland production and all of rainfed upland production. There is popular preference to glutinous rice by the farmers and are reluctant to use hybrid varieties of rice. Despite having high potential and traditional legacy for rice cultivation both in-house and for rice export, the fanners have not incorporated the adoption of hybrid varieties of rice and are still rooted to produce low quality rice variety using inferior quality of rice than compared to the other globally cultivated hybrid varieties of rice for rice export. This makes the standard of rice cultivation very low. Prior 1990 there was no national research network, nor were there accurate national soil maps and baseline meteorological information for many regions. It was only as late as 1990, with the support of International Rice Research Institute (IRRI) and funding from the Swiss

Development Cooperation, a national rice research programme and network were initiated by Laos.

3. *Deforestation in Uplands*: Soil erosion and fertility losses are rampant in upland rice (about 36 per cent of total area) cultivation. This is chiefly due to unabated deforestation. The forest cover of Laos is hence exponentially reducing rapidly, causing land slides (during the monsoon region) which misbalances the agricultural cultivable upland. Deforestation also causes periodic flooding of Mekong River (in the excessive rainfall period of July and August). This flooding causes much damage to the agricultural cultivable land. Often the floods get aggravated by two occurrences of tropical storms "Irwin" and "Loise" which pass near Laos, touching the Upper Basin of the Mekong River and the catchment areas of the Nam Ngum River. Severe flooding caused by these rivers usually affects around 10,000 villages and estimates reveal that annually around 62,348 hectares of rice fields are totally destroyed on average.
4. *Inadequate Agricultural Infrastructure*: The status of Laos agriculture is in a very underdeveloped and disorganized state as displayed by the poor infrastructural facilities and lack of governmental initiative in generating adequate credits and input supply for farmers such as—fertilizer and other agro-chemicals needed to enhance agricultural production. Problems of rainfed production are also aggravated by the relatively sandy and infertile soils that prevail in most of the rainfed lowlands. Fertilizer use is limited as organic fertilizer is usually applied only in seedbeds, while inorganic fertilizer use is limited to the relatively small area of irrigated production. This overall reduces the optimal production of crops and crop losses due to weeds, insects and crop diseases. Phosphorus and Potassium deficiency appears to be general in almost all areas of rainfed lowland cultivation. Between 10 per cent and 15 per cent of the rainfed lowlands have salinity problems. Appropriate fertilizer formulations and recommendations are not yet available. The lack of a developed marketing infrastructure system also indicates in-effective extension support by the state to agricultural developmental initiatives, as most farmers do not have ready access to fertilizer or credit to purchase it. The farmers find

themselves incapacitated due to price fluctuations caused either by drought, flood or any other environmental disasters and other market parameters that are globally instigated.

5. *Small Farm Size*: Farm sizes in Laos are very small and unconsolidated partly due to the state of impoverished fanners who feel orphaned due to lack of policy initiatives taken by the state and also due to the UXO contamination that prohibits the farmers to take measures for land consolidation voluntarily. However as Laos is gearing up with UXO eradication efforts, the scope of apt and effective use of agricultural land becomes vital while UXO eradication efforts become mandatory for land consolidation and generation of confidence among farmers in the otherwise war ravaged Laos.

### *Land Use Debacles and UXO Contamination*

Figure 5 depicts only 4 per cent of arable land provoking attention to understand the agricultural debacles of Laos. Laos occupies about 55 per cent of forested area containing mixed, deciduous, and evergreen forests apart from re-growth of plantations, evergreen and dry wood (scrubland) including bamboo with remaining land utilized for other purposes. 35 per cent of the land is littered with "Unexploded Ordinance (UXO)".[9]

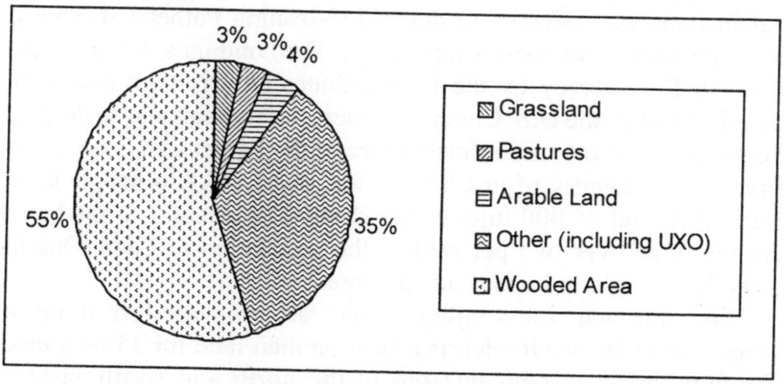


**Fig. 5:** Laos: Land Use (2004).

Grasslands and Pastures account for about 3 per cent of the land use in Laos sustaining livestock. In order to understand the imbalanced land use pattern in Laos, its important to examine the debacles it poses such that the prospects can be well underlined for achieving sustainability and effective resource management. (Fig. 1 shows the provinces effected by UXO and its contamination levels). During the American war on Vietnam from 1964-73, Laos was embroiled in the war in neighbouring Vietnam and as a result, it became one of the most heavily bombed countries in the world. "UXO are lethal weapons/ordnance dispensed from aircraft, missile, rocket, artillery projectile that break open in mid air releasing hundreds of small submunition (bombies)[10] over a wide area that fail to explode or detonate as designated".[11] Most accidents in Laos are caused by bombies; unlike landmines, these are designed to kill rather than injure. "Between 1964 and 1973, the United States dropped more than two million tons of bombs on Laos during 580,000 bombing missions-exceeding the total amount dropped on Germany and Japan during all of World War II—in an effort to fend off communist insurgents in South-east Asia. These bombing missions flown over Laos equaled one bombing mission every eight minutes 'round the clock, for nine full years". In Xieng Khouang Province, one of the most heavily bombed areas, an estimated 300,000 tons of bombs were dropped, of which an estimated 20 to 30 per cent did not detonate, equalling more than two tons per inhabitant. Today, Laos's ranks as the most bombed country (per capita) in all of warfare history for bombings included 90 million cluster bombs.[12] Specifically, the air war in Laos was targeted to defeat left-leaning Pathet Lao and to destroy North Vietnamese supply lines. The bombings in Laos were conducted in secrecy by the United States Government and were unauthorized by the U.S. Congress, where the strategy was to destroy the social and economic infrastructure of the Pathet Lao communist forces. "In a country of just 5.5 million, nearly 40,000 people have been killed and 65,000 injured by UXOs. This equates to the harm and injury of over two per cent of the population in Laos. One in every 380 people in Laos is an amputee."[13]

After the war, Laos villagers and farmers were left alone to struggle with the deadly debris of war on their land for 35-40 years. The bombing was concentrated in the north and south-eastern provinces of the country affecting 13 out of 18 provinces in Laos. It was in 1994 by the Mennonite Central Committee, the Mines Advisory Group, and the Lao Government initiated a UXO clearance programme and got attention from United Nations through the 1997 Ottawa Mine Ban Treaty. Currently, the Lao government is undertaking the UXO clearance, training and community awareness activities which are being conducted in nine provinces by six international mine clearance organizations and military advisers in order to achieve ***UXO-free agricultural land.***

The UXO Lao National Unexploded Ordinance Programme was established in 1996, which took up the responsibility to achieve an UXO-free land and enable appropriate land use in Laos. Figure 6 illustrates the level one 1996-1997—survey report on area under UXO contamination in Laos.

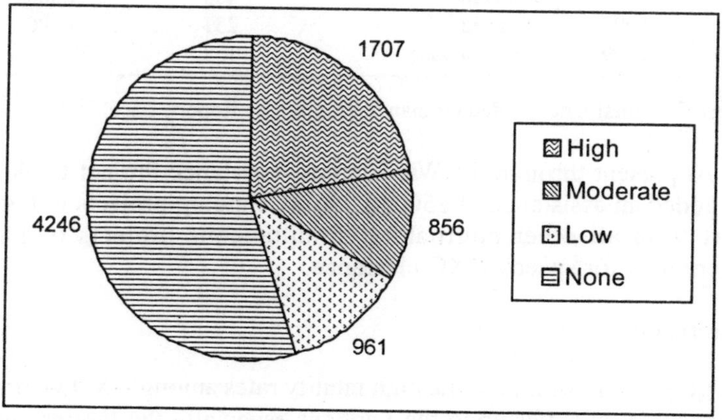


**Fig. 6:** Laos: UXO Contamination (Number of villages: 1997).

High contamination is identified where UXO contamination is in village centre or road. Moderate UXO contamination is identified in that of contamination in agricultural fields only and low UXO contamination is in grazing land or forest. This is designated with the level of land use and the number of accidents under UXO contaminated area. The 1996-97 Level One UXO survey coverage was done in 7700 villages (out of 11000), 93 of the 135 districts and 15 of the 18 provinces. The following table depicts the UXO Clearance status as of 2004 under the UXO Lao National Unexploded Ordnance Programme spreading UXO awareness in a total of 4532 villages and clearing 4392 hectares of the 9 heavily UXO contaminated active provinces of Laos, depicting steady progress despite the fact that the pace and nature of clearance is cumbersome and terribly lethal.

UXO LAO: National Unexploded Ordnance Programme: UXO Clearance Status 2004.

| *Sl. No.* | *Year when work started* | *District (Active Provinces)* | *Area cleared (hectares)* | *Disposed UXO items* |
|---|---|---|---|---|
| 1. | 1996 | Xieng Khonang | 1332 | 168,000 |
| 2. | 1996 | Houphan | 624 | 37,117 |
| 3. | 1997 | Savannahkhet | 646 | 116,000 |
| 4. | 1997 | Champassak | 274 | 48,000 |
| 5. | 1997 | Sekong | 203 | 15,700 |
| 6. | 1998 | Luang Prabang | 541 | 16,465 |
| 7. | 1998 | Saravan | 409 | 74,000 |
| 8. | 1998 | Attapeu | 231 | 26,000 |
| 9. | 1999 | Khanmouane | 132 | 26,000 |

*Source*: National Unexploded Ordnance Programme Report, Laos, 2004).

At present through the War Victims Assistance Project USAID has added an assistance of $500,000 to its ongoing projects in Laos apart from all other multilateral simultaneous projects of Lao government for speedy UXO clearance.

## Conclusion

The gory nature of injury, the high fatality rates among UXO victims is *unnecessary suffering* to the Laotians especially the farmers and poses as a grave human debacle. The lethality of cluster bombs' with its characteristics like small size, detonation by contact with persons, hidden, lethal over decades makes it inherently indiscriminate, be it people, farmers, children and livestock. It is mainly this lethality of UXO contamination that the agricultural land use debacles continue to persist in Laos and agricultural development is retarded. It is important to understand that in addition to reducing the human toll of deaths and injuries from explosions, UXO clearance helps reduce poverty by making more land available for farming and establishing new infrastructure as an imperative for development in an agriculturally rich potential country of Laos.

## NOTES

1. Geographical position in extreme coordinates.
2. Lao Embassy, Washington D.C., USA 2004.

3. Association for South-east Nations' or Asean was established on 8 August, 1967 in Bangkok. It's a 10-member country group comprising of Brunei Darussalam, Cambodia, Indonesia, Laos (joined on 23 July, 1997), Malaysia, Myanmar, Philippines, Singapore, Thailand and Vietnam.
4. C1A World Fact-book 2004.
5. Lao Export Promotion Division 2002.
6. *Ibid.*
7. Food and Agricultural Organization (FAO) estimates.
8. *Ibid.*
9. UXO is commonly referred term by mine action committee. Mines Action Canada, 1994 and does not include mines (designed to explode when activated by a person or vehicle) or abandoned munitions (items that have not been fully prepared tor use or not used at all).
10 Bombies: is the Lao name for anti-personnel Bomb Live Units (BLU 26). Most of the aerial bombardment consisted of anti-personnel cluster bombs filled with bomblet sub-munitions, which have a "spin-to-arm" mechanism and were intended to explode on impact or shortly-after impact. Each cluster bomb container was filled with upto 670 individual bomblets.
11. Identified as per "1997 Ottawa Convention on the Prohibition of the Use, Stockpiling, Production and Transfer or Anti-Personnel Mines and on Their Destruction" or "Mine Ban Treaty".
12. *Ibid.* Mines Advisory Group, 1995 Data, Canada.
13. Landmine Monitor 2003.

## REFERENCES

Branfman, Fred. (1993): *Voices from the Plain of Jars: Life Under an Air War*, New York: Harper & Row, 1972.

Castle, Timothy N. (1993): A War In the Shadow of Vietnam: U.S. Military Aid to The Royal Lao Government, 1955-1975. New York: Columbia University Press. 19.

Lao UXO Work Plans (1999): New Jersey: Zed Books.

Chapman, EC, Bounthong Bouahom and Peter Kurt Hansen (editors) (1997): Upland Farming Systems in the Lao (PDR)—Problems and Opportunities for Livestock. *Proceedings of an International Workshop held in Vientiane*, Laos 18-23 May.

Evans Grant (1988): *Agrarian Change in Communist Laos,* Institute of South-east Asian Studies. Singapore.

Izikowitz, Karl Gustav (1951): *Lamet: Hill Peasants in French Indochina.* (Ethrografiska Studier No. 17) Goteborg, Sweden: Etnografiska Museet.

Littauer, Ralph and Norman Uphoff (1977): *Air War in Indochina.* Air War Study Group, Comell University. Boston: Beacon Press, 1972. Mines Advisory Group, Data-Gathering team report for 1995, Moung Pek and Moung Kham,

Xieng Khouang, in Ann Martin, "New Batch of UXO Project Materials," MCC Mar. 96 Memo.

Prokosch, Eric (1995): *The Technology of Killing: A Military and Political History of Cluster Weapons.* London an Draft Agreement for the UNDP Trust Fund Nationwide Unexploded Ordnance Clearance Initiative Lao (PDR), 19 April.

Shifting Cultivation Project (1992): *Agriculture Programme 1990-1991, Technical Report* (Annex 2) Vientiane: World Conservation Union.

"The Initial Failure of Agriculture Collectivization in Laos," *Asia Quarterly* [Brussels], 4, 1980.

**Websites**

www.mmc.org
www.fao.org
www.uxolao.org

# PART III

# Urban Land Use Dynamics and Its Impacts

# 10

# Dynamics of Land Transformation in Relation to Distance

## *A Case Study of Growing Urban Center*

Shahab Fazal

Just after the millennium another great human milestone has passed, for the first time in history, a majority of the world's six billion people live in cities. It has been further estimated that between the years 2000 and 2025, the world's urban population will double from 2.4 billion to 5 billion and city dwellers will rise to over 60 per cent of the world's total population. Significantly, this increase would be more pronounced among developing countries.

India although known as land of villages but here also a steady growth in the size, proportion of the urban population as well as the number of urban centers has been recorded. India's urban population in the last fifty years has more than tripled at an average rate of 3.09 per cent per annum. The future projections indicate that it would reach up to 650 million by the year 2025 (Fazal, 2001).

In India, urban population is increasing significantly through migration of population to urban centers from rural areas. The main reason for this migration is the lack of employment opportunities in rural areas. Still, the main occupation in rural areas of India is agriculture, which is unable to bear the burden of rapidly increasing population. The agricultural land remains as the same but due to natural growth, the dependent population is increasing, leading to high land-man ratio (Fazal, 1998). The main basis of urbanization is economic change and in particular the growth of secondary and tertiary occupation in urban areas (Fazal, 2001). But there is vast difference in urbanization process especially internal structure and pattern of reorganization of land in India and the urban centers of the developed countries. Cities of developed countries are characterized by different land use dominated by specific class, Indian cities (developing country) tend to have different aerial blocks with several activities in each (Surekha, 1988). This results in division of a large city into a smaller entities, which are mostly self-contained and loosely coordinated with each other (Fazal, 2001). Several studies have highlighted this fact that the character of urban land use pattern in Indian cities is very different from those in western cities, but have also observed a degree of homogeneity within the Indian cities (e.g. Yadav, 1988). Breese stated that the characteristic feature of Indian urban land use is its high degree of mixing, often to the extent that it is difficult to discern any pattern. Dattatari, Misra and Arunachalam all have indicated towards indiscriminate land use in Indian cities. Similarly, studies also showed that in smaller Indian cities and towns, the inter-mixing of land use is even more pronounced (Fazal, 2001).

The city and its environs offer an excellent case study of land use issues. Choice of land use is made by the individual land user, operating with in broad physical, social, cultural and economic influences. That is why, contrast exist in the structures of land use in different countries (economies). Widespread concern is felt regarding the urban expansion and reorganization of land in changing socio-economic scenario in India. Here centrality is an important aspect in cities, their evolution and development starts from central point and all the major activities revolve around it. In fact the entire city traditionally is governed by this central part. Theoretically also, a direct relationship exists between land use, its location and land value, it also points towards elements of zonation in land use. The present study tries to evaluate the urban expansion, land reorganization in the city especially in relation with distance.

## Study Area

The study area lies between 29° 55′ to 30° north latitude and 77° 30′ to 77° 45′ east longitude. It is situated in Western part of Uttar Pradesh (India), in the fertile tract of Upper Ganga-Yamuna doab (inter-riverine plain). Saharanpur City is the district headquarters and second largest city in Meerut division. It is well connected by several state road highways to the important cities. City is having nodal location between the hilly districts of Uttaranchal (in the north) and plain region (in the south) of Uttar Pradesh (Fig. 1). The city is spread over 4,030 hectares (Municipal boundary). But since it has expanded beyond this limit so we have also extended our study area, which deals with 7,745 hectares. The present city is divided into two parts by the railway line, the northern part is the older part while southern part is newer, where most of the institutional and industrial development has taken place. The population of the city is 0.57 millions according to 2001census. The city has the population density of 239 persons per hectare.

## Data and Methodology

The present study is based on secondary data including the remotely sensed data (satellite imagery). This data along with extensive field checks and surveys provided accurate and up-to-date data. The advantages of these data are many, because urban areas are very dynamic and complex in nature and conventional data source were unable to cope with these data. The information was far too much and they too get changed in a very short period of time. On the other hand their collection took a very long period. So remote sensing is a strong tool for urban studies. Now IRS 1C is providing 5.8 metres resolution. Which makes the urban studies even more fruitful. The study is spread over a time span of twenty years (1982-2002).The urban sprawl mapping was done with the help of Saharanpur city Guide map, prepared by Survey of India (1982) on 1:20,000 scale, IRS 1C geo-coded panchromatic satellite analogue imagery on 1:25,000 scale, acquired in 2002. The minimum delineation unit for both the maps was 0.5 cm to 0.5 cm. The satellite imagery was first visually interpreted with the help of separate interpretation keys to identify various land use classes. Some of the land use classes could not be distinguished by visual interpretation of satellite imagery. Especially, the differentiation between residential and commercial land use classes. These classes were identified during the field checks and demarcated afterwards. Subsequently, the details of these final maps were transferred on Base map, bringing both the maps on same scale. Thereafter, these maps were digitized for analysis. Further the study was done in G.I.S. environment because the use of GIS techniques, coupled with remote sensing would be essential elements for the rapid production of accurate and up-to-date databases.

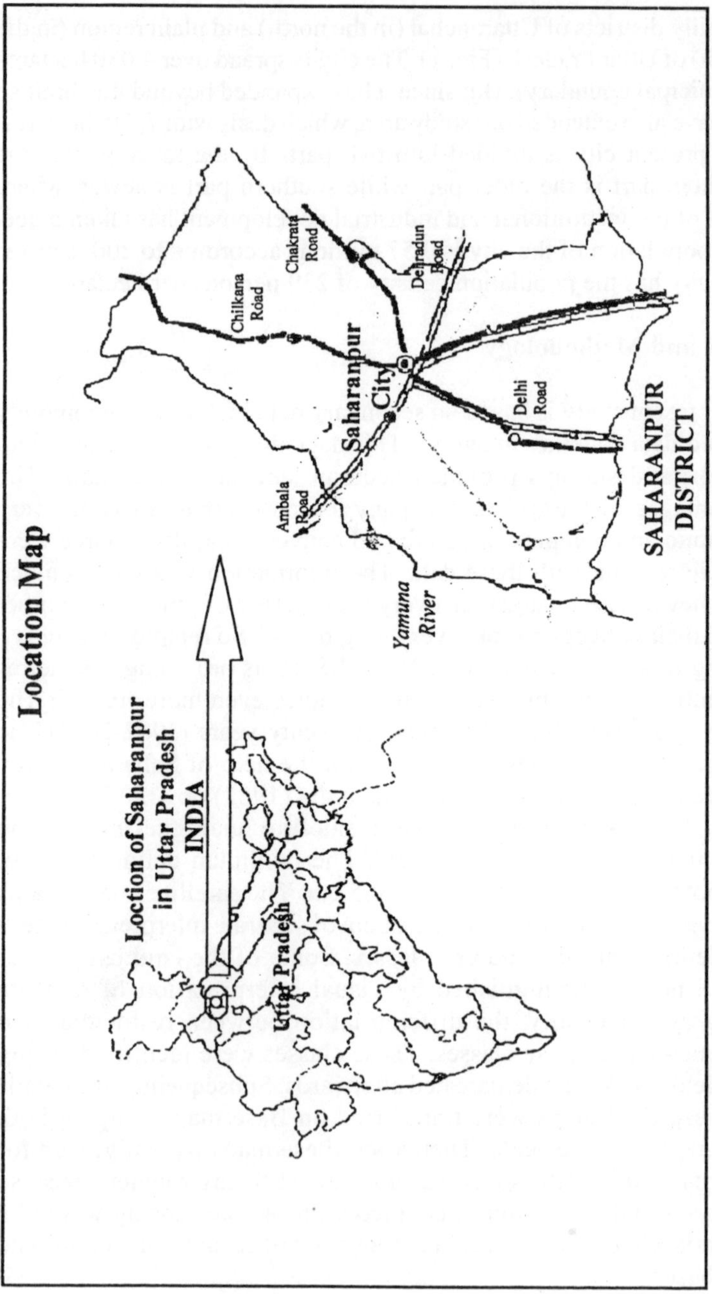


**Fig. 1:** Location Map of Saharanpur District, Uttar Pradesh

The inherent flexibility of GIS, techniques can make important contribution for improving the speed and qualitative performance of urban and regional planning. GIS techniques allow quite complex quantitative and spatial relationships to be made and displayed in graphical form and improve the clarity of the planning and decision making process. The present work has been carried out using ILWIS software (The Integrated Land and Water Information System) 2.2 windows version developed by ITC, The Netherlands.

## Discussion

Traditionally Saharanpur has agro-based economy. It lies in the fertile land between two master streams of India, the Ganga and the Yamuna. Initially the development by Saharanpur City was also because of its agrarian background. It had well-developed market for agro-based products. It also played the role of service centre for nearby towns and hilly region of Uttaranchal. But in the last 25 years several industries have also come up and changed the fabric of the city's economy. The city has well known paper, straw board, tobacco and cigarette industries besides these big industries several medium level agro-based, motor transport, hosiery industries are flourishing in the city. The city has also excelled in wood sculpture and its products are exported to several countries. That is why the city and its surroundings have under gone radical changes.

### *Urban Land Use of Saharanpur City*

Urban areas and its environs offer an excellent case-study of land use issues. Land use pattern are usually more complex and dynamic than in the deeper countryside. Saharanpur city is named after one Sufi Saint Shah Harun Chisti. It was ruled by Mughals, but the development of city took place only under British's who in 1903 made this city as district headquarter in 1926 when the city received electric supply its economic development got a boost and city started to develop at a faster speed. After independence a good number of refugees came to this place and settled here. The present city is divided into two parts by the railway line, the northern part is the older part while southern part is newer, where most of the institutional and industrial development has taken place.

Although the municipal area of Saharanpur city is only about 4,030 hectare but due to the objectives of our study we are dealing with 7,745 hectares of land. It includes the changes and urban expansion of the city, which have occurred in the city during the study period. From the perusal of Table 1 and Fig. 2, it is clearly evident that land use of Saharanpur city is not very well planned. With time, the older part of the city has become more and more congested while on the other hand the city has expanded outward in a haphazard manner and swallowed the agricultural land in the fringe area. The urban land use of Saharanpur city is dominated by residential class, which occupies more than 26.77 per cent of the study area. Most of this residential area is unplanned having very little open spaces and poor basic infrastructure facilities. The second largest area is occupied by vacant land (10.69 per cent), which is the outcome of rapid transformation of occupational structure due to emergence of secondary and tertiary activities. The main reason for the increase of vacant land in the city was that initially when industrialization took place, there was shortage of open land for construction. This demand has led to increase in the land value and owners of land started to get more money instantly, which they were not getting in agricultural sector. So in anticipation more and more people started to offer their land for construction resulting in phenomenal transformation of land from agricultural uses to vacant land. But later this demand due to stagnation and saturation did not subsidize, resulting in increase of vacant land, which is neither used for urban development, nor is useful for agricultural sector. Similarly, area under plantation (tree crop) is also spread over 5.92 per cent of the study area. This again is mainly due to two reasons, firstly, rich people are protecting their lands (in city area) from encroachment by planting trees and they are waiting for increase in land value. Secondly, it is also quite remunerative as well as assured market in the city itself, where it is used in paper industry. These factors helped in increase of area under plantation. It is also favoured than cultivation of crops in and around city region because it needs lesser care and protection. The city has 216 hectares of land under industries, which is quite high by traditional Indian standard. Commercial area in the city occupies 125 hectares of land. These areas are mainly in the central part of the city and that too along the major roads. Significantly, area under city's recreational, utilities and services (categorized as others) covers almost negligible area of study area. This indicates that, not much attention is paid for developing these facilities despite the substantial increase in population. The city also has a good share of land (5.2 per cent) under remount depot (restricted area) used for defense purposes.

The salient characteristics, which emerges from Table 1 and Figure 2 are as followed:

**Table 1:** Saharanpur City, Area Under Different Land Use Classes (1982-2002).

| *Land Use/ Year* | *1982* | | | *2002* | | | *Percentage Change 1982-2002* |
|---|---|---|---|---|---|---|---|
| | *Area* | *Proportion of total land (in percentages)* | *Proportion of developed land (in percentages)* | *Area* | *Proportion of total land (in percentages)* | *Proportion of developed land (in percentages)* | |
| Residential | 1084 | 13.99 | 52,88 | 2073 | 26.77 | 64.74 | 91.24 |
| Commercial | 44 | 0.56 | 2.14 | 125 | 1.61 | 3.9 | 184.09 |
| Institutional | 61 | 0.79 | 2.97 | 63 | 0.81 | 1.97 | 3.27 |
| Industrial | 122 | 1.58 | 5.95 | 216 | 2.79 | 6.75 | 77.04 |
| Agricultural Land | 4998 | 64.53 | – | 3257 | 42.05 | – | –34.84 |
| Tree Plantation | 375 | 4.84 | – | 459 | 5.92 | – | 22.4 |
| Vacant Land | 322 | 4.17 | – | 827 | 10.69 | – | 156.8 |
| Restricted Area | 402 | 5.19 | 19.61 | 402 | 5.19 | 12.55 | 00 |
| Others | 337 | 4.35 | 16.44 | 323 | 4.17 | 10.08 | –4.16 |
| **Total** | **7745** | **100** | | **7745** | **100** | | |

The city is expanding mainly towards north and south-west direction. The urban expansion towards north is mainly by industries (rice mills). This expansion has incurred loss to plantation, which earlier existed. The expansion towards south direction is somewhat limited because of unfavourable factors like location of rifle range, high-tension power line etc. Similarly, the city growth towards east has been restricted by Remount Depot (Restricted area). Recently rapid urban expansion is taking place along Delhi road in south-west direction. Some of the major industries like Cable netting, Rubber, Plastic and Cracker factory etc. have come up. Some unplanned residential area surrounds these industries. Similarly, in south-east direction along the railway track some unplanned residential area has come up. It is extended up to Star Paper Mills. Most of these developments have taken place on vacant land. In the north-eastern direction, along the Dehradun road, a ribbon like urban development is taking place. Which is marked by some small industrial units along the Dehradun road and in between these built-up area, vacant land has developed. In the central part of the city along the main roads, there was a significant increase in commercial area. This increase was because of conversion of unplanned residential area to markets.

The land use pattern of Saharanpur city is different from the general land use pattern of western countries. The existence of a mono-economic system and an advanced transportation technology of the western cities had led to a maximum specialization of land use which has most frequently taken the form of concentric circles of relatively homogeneous economic usage. From the perusal of Figure 2, no such clear zonation characterizes the land use pattern of Saharanpur city. It is hard to generalize about the patterns of residence. The rapid growth of population of the city associated with the socio-economic changes, which are creating an emergent middle class. This has caused a proliferation of squatter settlements in the interstices and fringes of the city. At present Saharanpur city shows little conformity with the western cities where direct relationship is observed between social class and distance from the centre of the city with working class living closest to the city core and upper class in the city fringe. The pattern of residence in Saharanpur city includes a mixture of poor and rich elements of the city in both the core and outer areas. It represents some transitory phase, which is developing between the patterns of pre-industrialism and post-industrialism. There was only one distinct commercial zone in Saharanpur city. This zone has peculiar radial dimension along the main roads from the *Ghanta ghar* (located almost at the centre of the city). It is the central economic zone, which consists of shops and showrooms of reputed products ranging from machinery, consumer goods and warehouses etc. Other commercial areas are scattered, but located along the roads and their concentration declines with distance. These commercial areas are crowded and picture of congestion and traffic *jams* and to the western observer, this is a place of noise and colour. The industrial land use of the city resembles the western location in some way. The larger industrial units are located in the fringe area but the main difference from western cities is that it is not confined to specialized estates but are scattered. The other land use classes like vacant land and plantation (tree crop) are governed by local factors. The same trend is not necessarily found in

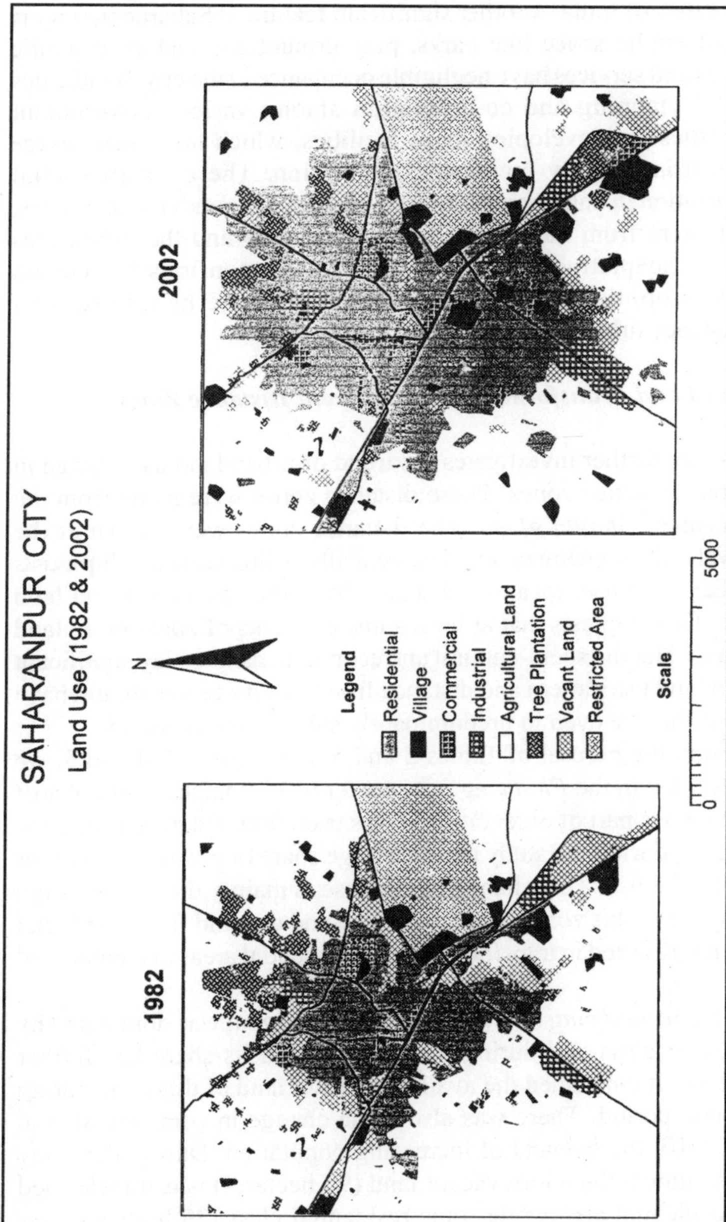


**Fig. 2:** Saharanpur city: Land use (1982 and 2002)

other cities of India. Another significant feature of Saharanpur city is lack of public space like parks, play ground etc. and even public utilities and services have negligible occupance in the city. It indicates lack of planning and co-ordination among various government authorities for developing these facilities, which are under severe stress from rapidly increasing population. The over all spatial organization in Saharanpur city represents indigenous characteristics, which stem from indigenous living conditions and the indigenous ways of adaptation with them. This pattern confirms the Gerald Breese's opinion for Indian urbanization where he termed it as subsistence urbanization (Breese, 1969).

### *Urban Land Transformation in Different Distance Zones*

The study further investigates the trend of urban land use change in different distance zones. These distance zones were made from the city center *(Ghanta ghar)*. The distance zones were drawn at the interval of 1 kilometre radius. Theoretically, a direct relationship exists between land use, location, distance from the city center and land value. This suggests that at least some elements of zonation in land use exist. But the socio-cultural and economic factors play significant role in land use pattern and distance from the city center should have some influence even in predominantly subsistence economy.

From the perusal of Tables 2 and 3 and Figures 3, 4 and 5, we observe that in the *First ring* (city center to 1 kilometre radius) most of the land is part of older city and was used for residential purposes. However, during the study period a large share of it was captured for commercial use. This change in land use is mainly due to very high land value in this zone. High land value and demand for commercial place has resulted in transformation of residential areas to commercial uses.

The *Second ring* (1-2 kilometre radius) again was dominated by residential class and during the study period its share has further increased. It consumed the available vacant land of this zone during the study period. There was also some change in commercial land use to fulfil the demand of increasing population. During the study period, almost the entire vacant land (95 hectares) was transformed into residential area while some residential class (29 hectares) was reorganized in favour of commercial uses.

**Table 2:** Saharanpur City, Land Use in Different Distance Zones (1982-2002).

| *Land Use* | *First Ring 0-1 K.M.* | | *Second Ring 1-2 K.M.* | | *Third Ring 2-3 K.M.* | | *Fourth Ring 3-4 K.M.* | | *Fifth Ring <4 K.M.* | |
|---|---|---|---|---|---|---|---|---|---|---|
| | *1982* | *2002* | *1982* | *2002* | *1982* | *2002* | *1982* | *2002* | *1982* | *2002* |
| Residential | 230 (69.69) | 199 (60.3) | 484 (57.41) | 564 (66.9) | 161 (9.64) | 542 (32.47) | 141 (6.51) | 589 (27.19) | 68 (2.48) | 179 (6.54) |
| Commercial | 18 (5.45) | 49 (14.85) | 13 (1.54) | 42 (4.98) | 9 (0.54) | 30 (1.79) | 4 (0.18) | 4 (0.18) | – | – |
| Institutional | 11 (3.33) | 11 (3.33) | 31 (3.68) | 31 (3.68) | 19 (1.14) | 21 (1.25) | – | – | – | – |
| Industrial | 4 (1.21) | 4 (1.21) | 30 (3.56) | 30 (3.56) | 28 (1.68) | 28 (1.68) | 29 (1.34) | 48 (2.22) | 31 (1.13) | 106 (3.87) |
| Agricultural Land | – | – | – | – | 805 (48.23) | 509 (30.49) | 1613 (74.47) | 748 (34.53) | 2580 (94.26) | 2000 (73.07) |
| Tree Plantation | – | – | 4 (0.47) | 4 (0.47) | 148 (8.87) | 86 (5.15) | 179 (8.26) | 209 (9.65) | 44 (1-6) | 160 (5.84) |
| Vacant Land | 13 (3.94) | 13 (3.94) | 113 (13.41) | 18 (2.13) | 178 (10.66) | 132 (7.91) | 10 (0.46) | 378 (17.45) | 8 (0.29) | 286 (10.44) |
| Restricted Area | – | – | 5 (0.59) | 5 (0.59) | 216 (12.94) | 216 (12.94) | 178 (8.22) | 178 (8.22) | 3 (0.11) | 3 (0.11) |
| Others | 54 (16.36) | 54 (16.36) | 163 (19.33) | 149 (17.67) | 105 (6.29) | 10 (6.29) | 12 (0.55) | 12 (0.55) | 3 (0.11) | 3 (0.11) |
| Total | 330 | 330 | 843 | 843 | 1669 | 1669 | 2166 | 2166 | 2737 | 2737 |

*Note*: 1. Area in Hectares.

2. Figures in brackets are percentages of total area of that particular ring.

*Source*:Based on Saharanpur City, Guide Map and IRS, 1c Satellite Imagery.

**Table 3:** Saharanpur City, Main Land Transformation in Different Distance Zones (1982-2002).

| *First Ring (0-1 KM.)* | | *Second Ring (1-2 KM.)* | | *Third Ring (2-3 KM.)* | | *Fourth Ring (3-4 KM.)* | | *Fifth Ring (<4 KM.)* | |
|---|---|---|---|---|---|---|---|---|---|
| Residential to Commercial | 31 | Residential to Commercial | 29 | Residential to Commercial | 21 | Vacant to Residential | 296 | Vacant to Residential | 27 |
| | | Vacant to Residential | 95 | Tree Plantation to Residential | 62 | Tree plantation to Residential | 107 | Tree Plantation to Residential | 39 |
| | | Others to Residential | | Agriculture to Residential | 185 | Agriculture to Residential | 45 | Agriculture to Residential | 45 |
| | | | | Agriculture to Vacant | 90 | Agriculture to Indust-rial | 19 | Agriculture to Industrial | 75 |
| | | | | Vacant To Residential | 155 | Agriculture to Tree Plantation | 59 | Vacant to Tree Plantation | 46 |
| | | | | Agriculture to Tree Plantation | 21 | Vacant to Tree Plantation | 108 | Agriculture to Tree Plantation | 109 |
| | | | | Vacant to Tree Plantation | 19 | Tree Plantation to Vacant | 30 | Agriculture to Vacant | 351 |
| | | | | Tree Plantation to Institutional | 2 | Agriculture to vacant | 742 | | |

*Note*:Area in Hectares.

*Source*:Based on Saharanpur City, Guide Map and IRS, 1c Satellite Imagery.

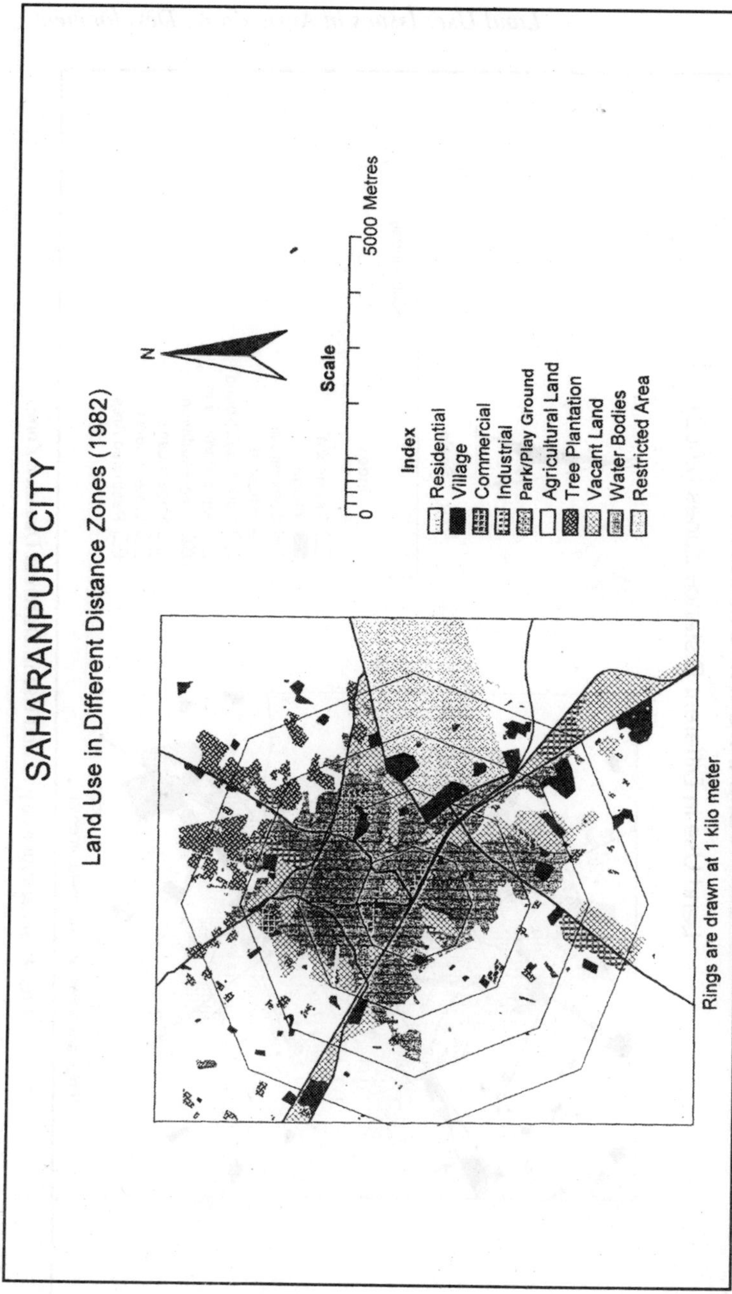


**Fig. 3:** Saharanpur: Land Use in Different Distance Zones

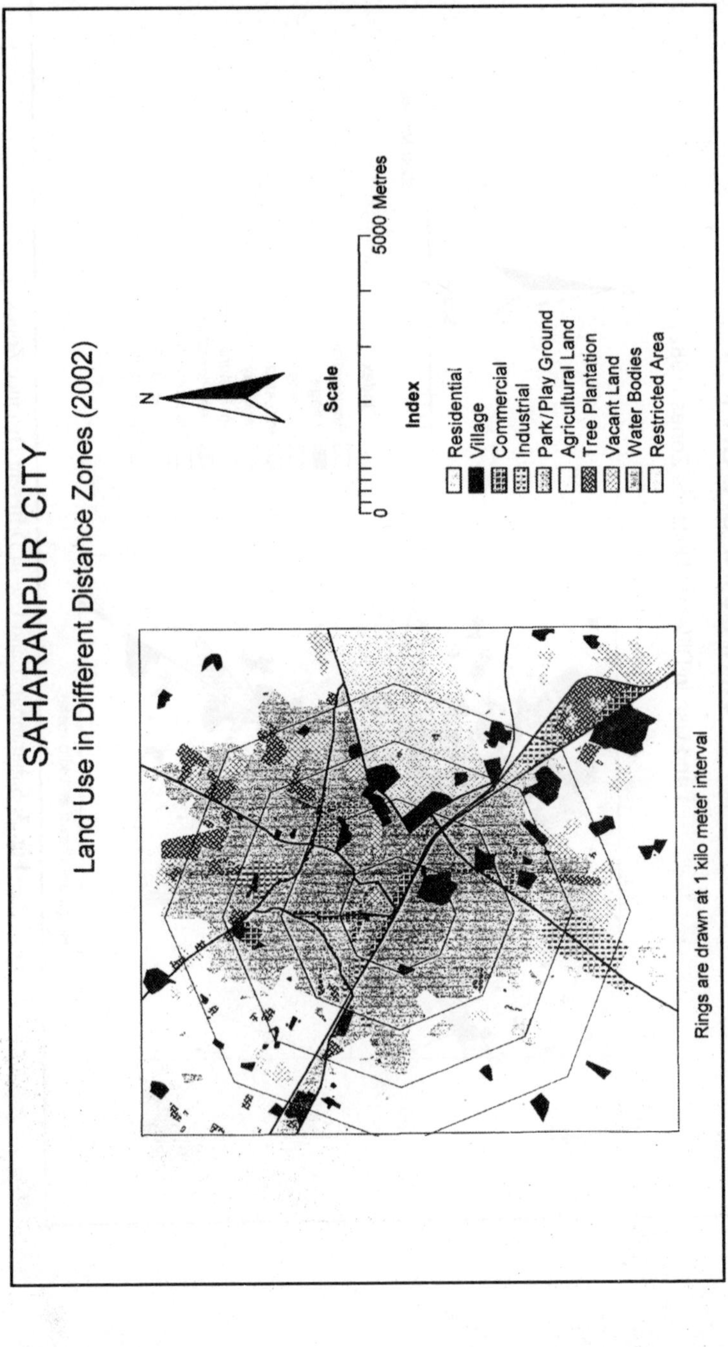


**Fig. 4:** Saharanpur city : Land Use in Different Distance Zones

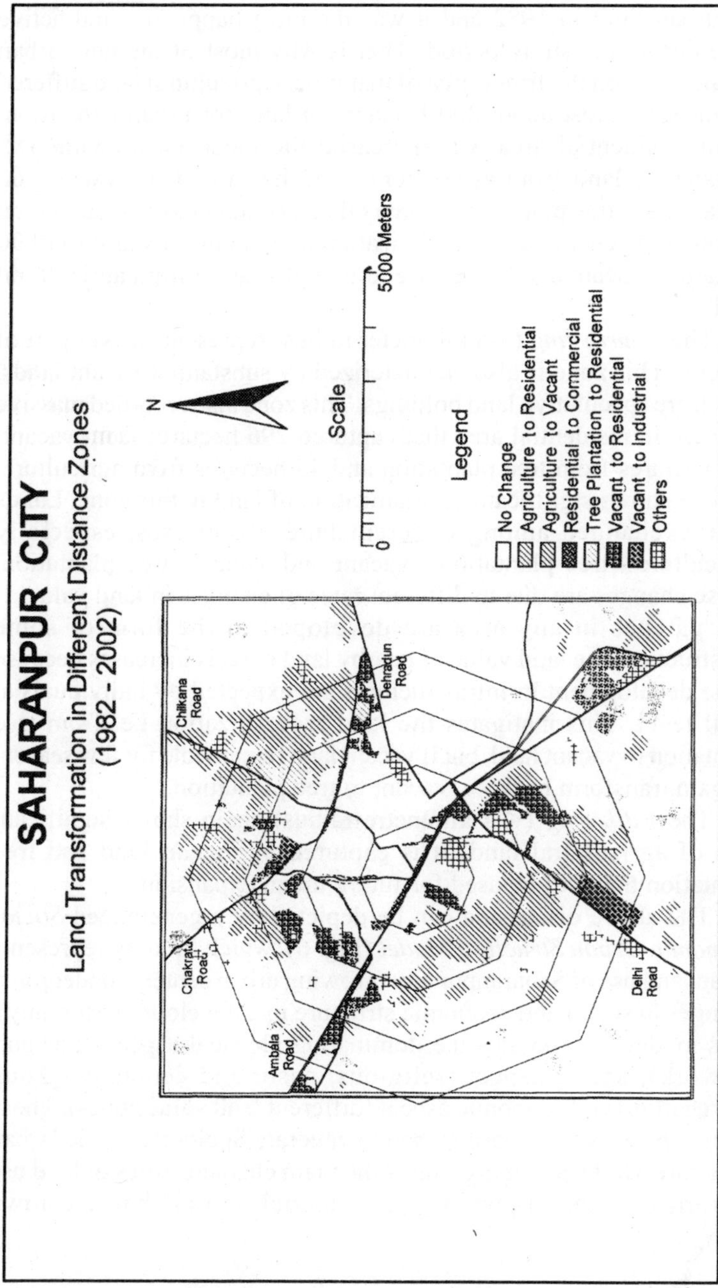


**Fig. 5:** Saharanpur city: Land Transformation in Different Distance Zone, 1982-2002.

The *Third ring* (2-3 kilometre radius) has recorded significant increase in residential class during the study period. This zone was the urban limit in 1982 and it was the most happening and active zone during the study period. That is why most of the new urban expansion is on the fringe area of that time. Agricultural land suffered the most and lost about 300 hectares of land for urban expansion. Again residential area was increased the most, swallowing 185 hectares of land from agricultural, 155 hectares from vacant, 62 hectare from tree plantation. Some other striking land reorganization follows the general land reorganization trend in the city, a shift of 90 hectare to vacant and 21 hectare to tree plantation from agricultural land.

The *Fourth ring* (3-4 kilometer radius) represents newer part of the city. This zone is also characterized by substantial vacant lands, which are speculative land holdings. This zone also recorded massive increase in residential area that captured 296 hectares from vacant, 107 hectares from tree plantation and 45 hectares from agriculture. Table 3 shows significant reorganization of land in this zone. Lands were exchanged among different land use classes, especially agriculture—tree plantation—vacant and vacant—tree plantation. These changes are due to different rates of changes in land value. If few patches in any area are developed in the form of some construction, the land value of nearby land parcels increases because some development in infrastructures are expected by individuals at local level. This instigates the land reorganization i.e. from tree plantation to vacant land, but if vacant land remain idle for few season, it again transform back i.e. vacant to tree plantation.

The *Fifth ring* (< 4 kilometre radius) again shows significant loss of agricultural land. It is captured by vacant land and tree plantation that will be used for future urban expansion.

The above discussion can be depicted by a generalized *Socio-economic Urban Structure Model* (Fig. 6), which broadly represents urban land use of Saharanpur city (growing urban center—undergoing changes in social and economic structure in a developing economy). This model incorporates the dominating physical aspect (transport network), social aspect (religious, caste and economic group segregation) and economic aspect (different land value zones). These three aspects were superimposed to generate Socio-economic Urban Structure Model and it highlights the main characteristics of land use pattern of Saharanpur city. The model shows that a Chowk

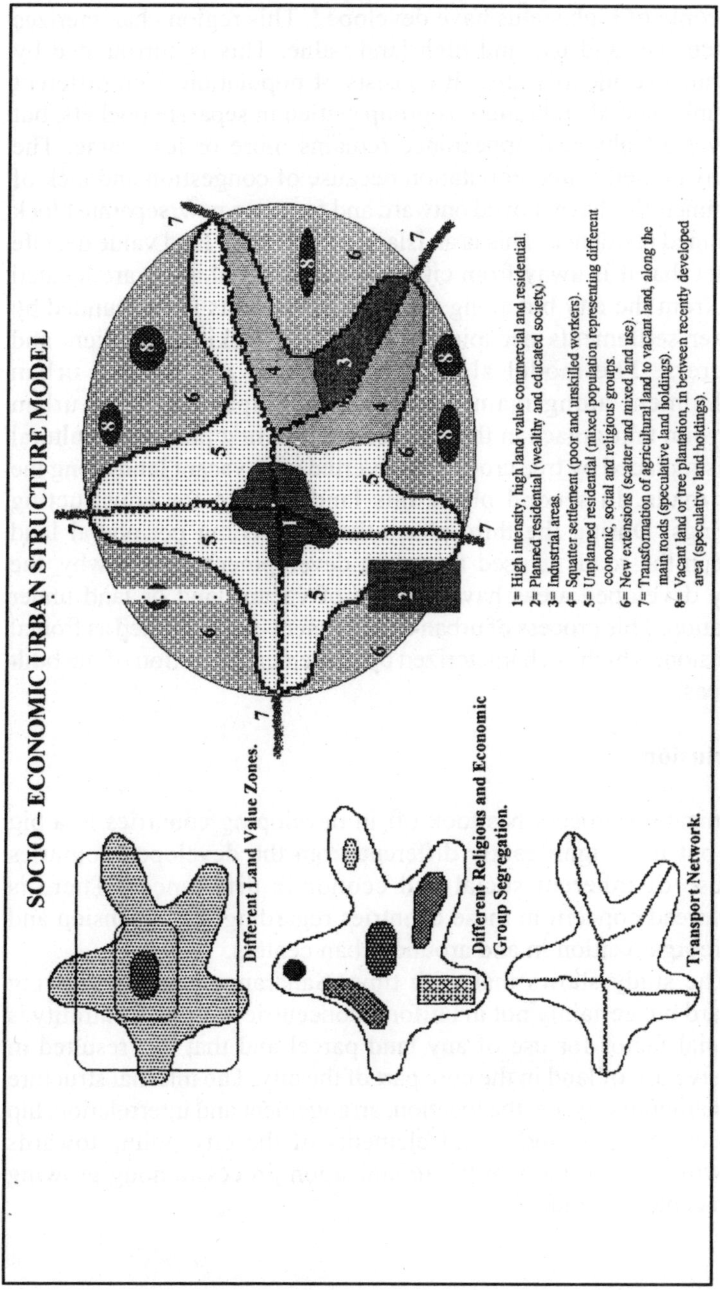


**Fig. 6:** Socio-Economic Urban Structure Model

(intersection of streets) occupies the city center (C.B.D.) around which shops and business establishments as well as the old houses of rich and people of high status have developed. This region characterizes by intensive land use and high land value. This is surrounded by unplanned residential area. It consists of population from different economic, social and religious group settled in separate pockets, but the over all physical appearance remains more or less same. The wealthy and educated population because of congestion and lack of basic amenities have moved outward and formed a new, separate block of planned residence. This is an island of very high land value despite the fact that it is away from city center. Industrial areas are located away from the city but along the main roads. This is surrounded by squatter settlements occupied by poor and unskilled workers and labourers. The model also points towards ribbon like urban development along main roads in Saharanpur city. The urban expansion takes place in three phases. In the first phase agricultural land is replaced by tree crops or converted in to vacant land along the main roads. In second phase this land is used for constructing buildings. During the third phase the intervening peripheral land between the roads is used for urban development, that is why, the newly developed areas have patches of vacant land or land under plantation. This process of urban development can be termed as frontal expansion, which is characterized by piecemeal accretion of the built up areas.

## Conclusion

Urbanization process has took off in developing countries in a big way, but it is significantly different from the developed countries because of inherent social and economic differences. There is widespread concern in these countries regarding the expansion and land reorganization in and around urban centers.

The study shows that with time, Saharanpur city is growing outward but certainly not in uniform concentric pattern. Centrality is a crucial factor for use of any land parcel and that has resulted in intensive use of land in the core part of the city. The internal structure of Saharanpur city, i.e. the location, arrangement and interrelationship between physical and social elements of the city points towards following crucial facts of the urbanization process among growing urban centers in India:

- The expansion of urban areas is swallowing fertile agricultural land, which is a very serious issue as the 'land is finite' and the converted land cannot be recovered back and is destroyed (for agricultural activity) for ever. This loss of agricultural land is even more serious in India because agriculture is the main occupation and country has very fertile land best suited for agricultural activity. As city expand, they not only displace agricultural land at the urban edge, but also cast a shadow ahead of them. Urban influences, in one form or another are felt on agricultural land much before bricks and mortar displaces.
- Vacant land in and around Saharanpur city has increased substantially. The main reason for its increase is the shortage of open land for construction. This demand has led to increase in the land value and owners of land started to get more money instantly, which they were not getting in agricultural sector. So in anticipation more and more people started to offer their land for construction resulting in phenomenal transformation of land from agricultural uses to vacant land.
- Similarly area under plantation (tree crop) has significant share in the study area. This is mainly due to two reasons, firstly, rich people are protecting their lands (in city area) from encroachment by planting trees and they are waiting for increase in land value. Secondly, it is also quite remunerative as well as assured market in the city itself, where it is used in paper industry. These factors helped in increase of area under plantation. It is also favoured than cultivation of crops in and around city region because it needs lesser care and protection.
- There was massive increase in residential area but not much change has occurred among utilities, services and recreational facilities etc., that indicates haphazard city expansion.
- Ribbon like urban development along main roads is a noticeable phenomenon in the city. This is because of higher land value along the roads due to better accessibility.
- The urban expansion is taking place in a peculiar way. In the first phase agricultural land is replaced by tree crops or converted into vacant land along the man roads. In second phase this land is used for constructing buildings. During the third phase the intervening peripheral land between the

roads are used for urban development. This process of urban development can be termed as frontal expansion, which is characterized by piecemeal accretion of the built up areas.

- The land transformation in Saharanpur city is clearly the outcome of the operation in the urban land market. It has many dimensions but the most important and relevant in present study was differential land value and land transformation in Saharanpur city which is basically the outcome of competitive bidding among various land use classes. However, in India social factors are still quite influential and they act as barrier in land transformation. That is why a large part of urban land remained frozen into distinct and mutually exclusive type of use and that too in the inner rings of the study area.
- The characteristic feature of urban land use in Saharanpur city (like most of the Indian cities) is its high mix and it is very difficult to make out any pattern out of it. Specialization of activities in a specific area, so common in the western cities, does not exist here. The spatial organization in the Indian towns and cities is distinguished by the diversified occupance of space. The Indian towns and cities have evolved this peculiar way of adaptation with space, is not based on economic interdependence and segregation.
- The Socio-economic Urban Structure Model very appropriately shows the land use of Saharanpur city as a Chowk (intersection of streets) occupies the city center (C.B.D.) around which shops and business establishments as well as the old houses of rich and people of high status have developed. This region characterizes by intensive land use and high land value. This is surrounded by unplanned residential area. It consists of population from different economic, social and religious group settled in separate pockets, but the over all physical appearance remains more or less same. The wealthy and educated population because of congestion and lack of basic amenities have moved outward and formed a new separate block of planned residence. This is an island of very high land value despite the fact that it is away from city center. Industrial areas are located away from the city but along the main roads. This is surrounded by squatter settlements occupied by poor and unskilled workers and labourers.

Lastly, we can say that urbanization pattern of Saharanpur city is characterized by poverty, acute shortage of housing and civic amenities, dismal filth and diversified occupance of space, which resembles like a large village. Here in lies the major departure of the urbanization pattern in Indian cities from the western cities. Urbanization may be an indicator of development in western countries but in India, especially in growing small towns it is difficult to accept. Here, the population is increasing rapidly, the demand for land also increases. The economy is heavily dependent upon the informal sector causing widespread informal hyper unwarranted urban expansion.

## REFERENCES

Alam, M. (1965): *Hyderabad-Secundrabad: A Study in Urban Geography*, Asia Publishing House, Hyderabad.

Borah, J. (1985): *Spatial Structure of Urban Influence in the Neighbouring Areas of Gauhati City*, Concept Publisher, New Delhi.

Bose, A. (1988): *India's Urbanization, 1901-2001*, Tata-McGraw Hill Publication, New Delhi.

Breese, Gerald (1969): *Urbanization in Newly Developing Countries*, Prentice-Hall of India Pvt. Ltd., New Delhi.

Burrough, P. A. (1968). *Principles of Geographical Information Systems for Land Resources Assessment*, Oxford Science Publication, U.K.

Deshpande, C.D. *et al.*, (1980): *Impact of a Metropolitan City on me Surrounding Region*, New Delhi.

Dirk, B. (1995): 'An Urban Community Strategy for Implementing Decentralization and Revolution Policy', *ITC Journal*, Netherlands.

Drabkin, H.D. (1977): *Land Policy and Urban Growth*, Pergamon Press, U.K.

Fazal, S. (1998). *Agricultural Production and Price Mechanism*, Chugh Publications, Allahabad.

Fazal, S. (2001): 'The Need For Preserving Farm Lands. A Case Study from a Predominantly Agrarian Economy', *Landscape and Urban Planning*, Vol. 55, No. 1, Elsevier Science Publication, U.K.

Fazal, S. (2001): 'Land Reorganization Study Along Major Roads. Land Use Policy', *Elsevier Science Publication*, Vol. 18, No. 2. United Kingdom.

Fazal, S. (2000) 'Urban Expansion and Loss of Agricultural Land. A Study in Indian Context Using Remote Sensing and G.I.S. Techniques'. *Environment and Urbanization*, Vol. 12. No.2, IIED, London.

Fazal, S. (2001): 'Application of Remote Sensing and GIS Techniques in Urban Sprawl and Land Use Mapping. A Case Study of A Growing Urban Centre in India', *Asian Profile*, Vol. 29. No. 1. Canada.

Fazal, S. (2001): 'Dynamics of Urban Land Transformation in Developing

Economy', *Indian Journal of Regional Sciences*, Indian Institute of Technology, Kharagpur, Vol. XXXIII, No. l.

Johnson. J M. (1984): *Sub-Urban Growth*, John Willey & Sons., London.

Misra, R.P. (1998): *Million Cities of India*, Vikas Publishing House Pvt. Ltd., New Delhi.

Prakasha Rao, V.L.S. (1983): *Urbanisation in India: Spatial Dimensions*, Concept Publishing Company, New Delhi.

Surekha, Pandit (1988): 'Urbanization the Indian Way' (ed.) Yadav, CS. *Perspectives in Urban Geography*, Volume 3, Concept Publishing Company, New Delhi.

Verma, S.S. (1995): *Urbanization and Regional Development in India*, Chugh Publications, Allahabad.

Yadav, C.S. (1988): *Comparative Urbanization. City Growth and Change*, Concept Publishing Company, New Delhi.

# 11

# Urban Land Use and its Impact on Groundwater Quality in the Delhi Metropolitan Region

Anju Singh and R.B. Singh

## Introduction

In recent years, water supply in India is being threatened due to the degradation in quality of groundwater. An uncontrolled disposal of urban waste into water bodies, open dumps and poorly designed landfills has caused groundwater contamination. This issue of groundwater pollution have become one of the most important lexicological and geo-environmental issues in India. As many critical regions depend on groundwater resources for various needs due to scarcity of surface water. In industrial and urban fringe zones of Delhi, the contamination of groundwater by industrial effluents and domestic waste is another serious challenge. The subsoil water in the area through which effluent channels from major industries flow, is permanently polluted and industries are releasing high concentration of toxic substances. The wells also in many residential areas are contaminated with nitrate, and detergents. The high content of fluoride in groundwater resources poses negative effects and is subject to severe health hazard in the surrounding region.

About 75-80 per cent human requirements are fulfilled by groundwater. Further groundwater accounts for more than 95 per cent of all (liquid) fresh water available on the earth at any given moment. In recent studies it was observed that there is an accumulation of inorganic substance from the high inputs of various chemical fertilizers, as a result the groundwater has showed increasing nitrates. World Health Organisation (WHO) has recommended that water supplies containing high levels of nitrate (100 mg of $NO_3$ per litre) should not be used for preparation of infant foods, alternative supplies having a low level of nitrate content, even to the extent of using boiled water has been recommended.

Groundwater pollution is emerging as an important environmental issue all over the world. The life support systems are increasingly threatened by both anthropogenic activity and waste mismanagement. In India, industrial contamination sources include wastes sites, storage and transport of hazardous substances, etc. Agricultural contamination sources include use of fertilizers, herbicides and pesticides. Urban air pollution also contributes contaminations by acid precipitation. The groundwater pollution is closely linked to land use activities. The various anthropogenic activities causes groundwater pollution issues, such as choice of building materials and waste handling methods.

Groundwater systems are highly sensitive to land and water pollution in areas of groundwater recharges but also to changes in flow condition. Sometimes, water table changes may cause pollution of groundwater. In recent days severe arsenic pollution of drinking water sources in India is attracting large attention due to health effects in terms of skin diseases. The management of such issues requires an integrated approach with public policies that support planning for the use and sustainable management of land and water resources. There is a need to develop close linkages between economic forces, land use and landscapes effects (Hilding-Rydevik and Johanson,1997).

## The Study Area: Delhi Metropolitan Region

Geographically, Delhi is situated along the western and eastern banks of the Yamuna River. Its altitude is 220 mtrs. above mean sea level and difference between the maximum and minimum levels in the cities about 60 mtrs. The entire territory is divided into five administrative blocks, viz. Alipur, Kanjhawala, Najafgarh, Shahdara and Mehrauli. Recently another city block was created. In the present study, two blocks, i.e. Alipur and Kanjhawala have been selected for detailed empirical study. The climate of the territory is semi-arid. The mean maximum temperature varies between 27.6°C and 40.2°C.

The mean minimum temperature ranges between 7.5°C to 22.2°C. The annual average rainfall is 754 mm which generally falls from June to September. The territory is made up of quartzite ridge extending from north and north-east (NNE) to south and south-west (SSW). The alluvial plain on the east of the ridge and sand dunes at the west are drained by the river Yamuna, which enters from the north-eastern direction and flows through the south-eastern area. The quartzite ridge is the prolongation of the Aravali group of rocks and enters from the southern border and ends in the North of Delhi on the west bank of Yamuna. The ridge forms the principal watershed in the area, the eastern drainage passing into the Yamuna and the western into the Najafgarh basin.

**Environmental Impact of Urban and Industrial Development**

Delhi, India's capital, ranks third in population among Indian cities and thirteenth among cities of the world. The growth of Delhi had been slow and gradual until India's independence in 1947. The industrial growth in Delhi got intensified only after India's Independence. From a small number of about 8,000 industrial establishments in 1950-51, the number increased to 90,000 in 1995. Government policies helped this rapid establishment of industries by providing facilities and incentives. This resulted in a mixed and even conflicting land use and causing deterioration in the quality of land and water.

The prime cause of poor sanitary conditions in many cities in India is due to the lack of facilities to collect wastewater and to dispose of after treatment. Data on wastewater generation and collection is less when compared to information on water supply. Owing to this, it is difficult to access the total pollution potential. The municipalities dispose of their treated or partly treated or untreated wastewater into natural drains joining rivers or used on land for irrigation or fodder cultivation or into the groundwater. The city of Delhi is only partially served by sewers. Even in the areas covered by sewers, all sources of waste water (including households) are not connected to the sewerage system. Usually one would expect 80 per cent of the water supplied to the community to return as wastewater. The daily flow of wastewater currently received and handed at the three wastewater treatment plants serving Metropolitan Delhi (Central Pollution Control Board, 1995).

Out of the total pollution loads generated from domestic wastewater, only a part terminate into the wastewater treatment plants and the remaining portion finds way, directly or indirectly (through seepage and overflow etc.) into the river Yamuna through the various open drains. Having no better basins for the assessment of the domestic wastewater and pollution load reaching various drains, it was assumed that roughly 50 per cent of the wastewater and pollution load reaching are carried away by sewers and the rest end up in drains. These are causing severe problem for groundwater pollution. The individual establishments in Delhi are primarily located in Western Delhi at Okhla Industrial Estate in Southern Delhi and the Trans-Yamuna area in Eastern Delhi (Table 1).

**Table 1:** Relative Contribution of Pollution Loads.

| *Drain* | *Item* | *Discharge KLD* | *Parameters of Pollution* | | |
|---|---|---|---|---|---|
| | | | *BOD Kg./Day* | *TDS Kg./Day* | *TSS Kg./Day* |
| Najafgarh | Domestic wastewater | 270000 | 725000 | 145000 | 72500 |
| | Industrial wastewater | 18100 | 5500 | 19800 | 6800 |
| | **Total** | **5500** | **78000** | **164800** | **79300** |
| | % Industry | 6.3 | 7.1 | 12.0 | 8.6 |
| Trans-Yamuna MCD | Domestic wastewater | 32000 | 5225 | 10450 | 5225 |
| | Industrial wastewater | 4300 | 1500 | 1000 | 1500 |
| | **Total** | **36300** | **6725** | **11450** | **6725** |
| | % Industry | 11.8 | 22.3 | 8.7 | 22.3 |
| Kalkaji | Domestic wastewater | 9000 | 2800 | 5600 | 2800 |
| | Industry wastewater | 400 | 400 | 300 | 400 |
| | **Total** | **9400** | **3200** | **5900** | **3200** |
| | % Industry | 4.2 | 4.8 | 6.7 | 4.8 |
| All Drains Combined | Domestic wastewater | 514560 | 147300 | 294600 | 47300 |
| | Industry wastewater | 22800 | 7400 | 21100 | 8700 |
| | **Total** | **537360** | **154700** | **315700** | **156000** |
| | % Industry | 4.2 | 4.8 | 6.7 | 5.6 |

*Source*: Based on Central Pollution Control Board, 1979; 1995.

## Groundwater Quality

In Alipur block, out of 21 stations only five stations, viz. Wazirabad, Bakhtawarpur, Holambikalan, Jindpur, and Hiranki show comparatively high salinity. Others indicate low salinity. A higher value of residual sodium carbonate (RSC) in irrigation water causes increase of sodium adsorption and thus the value greater than 5 meq/1 shows detrimental effect on crop growth because of sodium hazard. In case of anions, the concentration of chloride and sulphate to some extent are increased. The levels of nitrate in all the groundwater samples are within the permissible limit set for drinking water purposes.

Comparatively, Kanjhawala block shows more salinity, i.e. > 4,000 umhos/cm at Chandpur, Nithari and Tikri Kalan. In all other stations, the groundwater have either low or marginal salinity. The RSC values in some places such as Mundka, Bawana and Qutabgarh are more than 5 meq/1. This indicates that those areas are affected by sodium hazard (Central Board for the Prevention and Control of Water Pollution, 1985). Like Alipur block, the predominant caution in Kanjhawala is also sodium whereas a considerable number of stations contain chloride ion as dominant anion. The areas below the Rohtak road have high conductivity values. The stations, e.g. Khor Punjab, Bazidpur, Daryapur Kalan and Bawana may be grouped as moderate saline area (Tables 2 and 3).

The overall situation favours Alipur block for agricultural activities in comparison to Kanjhawala block. The groundwater from Kanjhawala block and upper Rohtak road may be considered as the enclosure of sodium hazard and high salinity. Joint balances of five block zones assumed on the basis of salinity in the two block indicate high salinity in Kanjhawala block than that of the Alipur block when groundwater of the saline enclosure zones of both the blocks are compared.

## Status of Heavy Metals in Groundwater

The trace elements surveyed in all those stations are lead, zinc, copper, nickel and cadmium. Other metals except lead in few well waters, were within the permissible level of drinking water. Data revealed that out of 30 locations in Alipur and Kanjhawala block (fifteen each), 3 locations in Alipur and 2 locations in Kanjhawala block were contaminated with calcium. None of the well water monitored in Alipur block indicated concentration of calcium exceeding drinking water standard for calcium in well water exceeding drinking water standards ranges between 11 and 25 μg/1. The calcium concentration of the well, which exceeded drinking water standard, is 17 and 11 μg/1, respectively. Other metals except lead in few well waters, were within the permissible level of drinking water. Out of 42 wells in both the blocks, the survey revealed that the level of lead in 6 wells, 3 in each blocks, exceeded the drinking water standard. These wells are Mukhmelpur, Holambi Kalan and Lampur, (Alipur block 142-408 μg/1) and Ladpur, Chandpur and Nithari (Kanjhawala block, 102-204 μg/1). The lead contamination in the well water of Chandpur were recorded as 110 and 204 μg/1 respectively in both the layers (Tables 2 and 3).

**Table 2:** Maximum Permissible Limits and Observed Value of Parameters in the Groundwater in Rural Delhi.

| *Parameters* | *Maximum Permissible Limits for drinking Water (μg/1)* | *Frequency Distribution of Actual Observed Values* | | | |
|---|---|---|---|---|---|
| | | *No. of samples out of 42 samples* | | *Concentration (μg/1)* | |
| | | *Alipur Block* | *Kanjhawala Block* | *Alipur Block* | *Kanjhawa Block* |
| Cadmium | 10 | 21 | 17 | 0-5 | 0-6 |
| | | | 4 | – | 11-25 |
| Lead | 100 | 18 | 18 | 0-58 | 0-96 |
| | | 3 | 3 | 142-428 | 102-204 |
| Chromium | 50 | 20 | 21 | 0-34 | 0-27 |
| | | 1 | | 50 | – |
| Zinc | 15,000 | 21 | 21 | 0-34 | 0-78 |
| Copper | 1,500 | 21 | 21 | 0-38 | 0-31 |

*Source*: Central Board for the Prevention and Control of Water Pollution (1985).

In terms of irrigation water quality the heavy metal concentration in all the groundwater stations of Alipur and Kanjhawala block were within the permissible limit particularly with respect fine textured soils. Among anions, nitrate shows the highest measurement of variability in both the blocks. The values of nitrate never exceeded the permissible limit (50 mg/1 as No. 3). The concentration of all the

**Table 3:** Areas having poor quality and polluted groundwater.

| *Block* | *Salinity* | *Nitrate* | *Fluoride* | *Heavy Metals* |
|---|---|---|---|---|
| Kanjhawala | Entire block saline | No serious problems except around few villages | Around Sultanpuri, Nangloi, Kanjhawala, Mundka | Mainly western part upto 40 mts. of depth (lead) |
| Najafgarh | Entire block saline except the drain and SW part | Present along with high fluoride (i) SW part | Except control part whole block is polluted | Mainly cadmium in central and western parts, lead in SE part |
| Shahdara | No serious problem | No serious problem | No serious problem | Highly polluted by lead, cadmium, chromium up to 40 mts. of depth |

*Source*: Based on Central Groundwater Board (1998) and Central Pollution Control Board (1998).

trace metals are well within the limit of drinking water standard. In Kanjhawala block almost all the parameters are highly correlated pair-wise except bicarbonate. Data reveals that conductivity and TDS have strong and positive correlation with other parameters except bicarbonate. In Alipur block, like Kanjhawala block, conductivity is well correlated followed by total dissolved solids, but in other cases the correlation are poor. As conductivity is highly correlated with other parameters mostly common ion, it is expected that the more or less similar results may be obtained in regard to other parameters too. It is, therefore, recommended that the present monitoring station with these seasons for a year, may be correct approach of designing monitoring network for groundwater. The concentration of all the trace metals are well within the limit of drinking water standard. However, some occasional negligible pulses are noticed in few stations of both the blocks. The coefficient of variations determined for ranking in regard to variability indicates that copper shows maximum variation and as such occupies the prime position in both the blocks whereas cadmium and lead show minimum variation. On the basis of detailed analysis of groundwater quality, Central Ground Water Authority (1998), New Delhi recently informed public that groundwater from shallow water bearing zones up to 30 metres of National Capital

Territory of Delhi has high concentration of Fluoride and Nitrate and is not fit for drinking purposes. It is advised that groundwater of the areas given in Table 4 should not be used for drinking purposes without proper treatment.

**Table 4:** High Concentration of Fluoride and Nitrate in Delhi Metropolitan Region.

| *Areas* | *Fluoride (mg/l)* | *Nitrate (mg/l)* | *Areas* | *Fluoride (mg/l)* | *Nitrate (mg/l)* |
|---|---|---|---|---|---|
| | | | **Block-Kanjhawala** | | |
| Bawana | 3.18 | – | Shahbad Dairy | 3.03 | – |
| Gherva | 2.21 | – | Tikri Kalan | 5.47 | – |
| Garhi Ringhala | 2.59 | – | Kanjhawala | 10.00 | 140 |
| Sultanpur Dabas | 2.6 | – | Korala | 1.53 | – |
| Mohammad Pur Mairi | 1.92 | 161 | Agar Nagar | 2.43 | – |
| Mundka | 2.29 | 119 | Kirari | 4.21 | 184 |
| Sultanpuri | 4.9 | 208 | Mangolpur Khurd | 6.1 | 114 |
| Mangolpur Kalan | 4.06 | 652 | Rajiv Nagar | 4.78 | – |
| Rithala | 5.1 | – | Punjab Khor | 3.48 | – |
| Qutabgarh | 5.26 | – | Katewra | 4.07 | – |
| Mungeghpur | 6.43 | – | Auchandi Border | 3.73 | – |
| Nijampur | – | 106 | Mangolpuri | – | 180 |
| Bhagey Vihar | – | 115 | Jat Khor | – | 390 |
| Krishna Vihar | – | 141 | | | |
| | | | **Block-Najafgarh** | | |
| Palam Gaon | 3.3 | 105 | Pappan Kalan | 3.65 | – |
| Bamnauli | 3.65 | – | Chawala | 1.8 | 130 |
| Kangan Hari | 1.9 | 300 | Daulat Pur | 2.05 | – |
| Shikarpur | 2.55 | – | Goela Khurd | 7.25 | – |
| Kharkhari Nahar | 1.6 | 203 | Nangoli | 3.8 | – |
| Ugersain Park | 2.4 | – | Kakrola | 3.05 | 379 |
| Mitraon | 5.47 | – | Pindawala Kalan | 2.21 | 361 |
| Dhansa | 3.84 | 377 | Ujwa | 2.1 | 508 |
| Jharoda Kalan | 7.3 | 208 | Dichaun | 3.69 | 178 |
| Neejwal | 5.08 | 117 | Jafarpur | 6.9 | – |
| Qazipur | 11.0 | 389 | Kair | – | 264 |
| Isapur | – | 106 | | | |
| | | | **Block-Alipur** | | |
| Shahbad | 2.14 | – | Puthkhurd | 3.48 | 266 |
| Pehladpur | 5.42 | – | Palla | – | 277 |
| Bakoli | – | 113 | Tikri Khurd | 3.2 | – |
| Mamurpur | – | 236 | | | |
| | | | **Block-City** | | |
| Munirka DDA Flats | 2.2 | – | Sant Nagar | 4.4 | – |
| India Gate Nursery | 2.54 | 121 | Delhi Zoo | 2.97 | 1589 |

*(Contd.)*

*(Contd.)*

| *Areas* | *Fluoride (mg/l)* | *Nitrate (mg/l)* | *Areas* | *Fluoride (mg/l)* | *Nitrate (mg/l)* |
|---|---|---|---|---|---|
| Nizamuddin | 4.83 | – | Jamia Millia | 2.46 | 1994 |
| Ashram Chowk | 2.3 | – | UPSC | 1.75 | – |
| Vikas Puri | 1.85 | – | Rajouri Garden | 4.25 | – |
| Meera Bagh | 2.05 | – | Vasant Enclave | 2.43 | 133 |
| Hastal | 1.85 | 171 | Parsi Dharamshala | 2.03 | 100 |
| Majnu Ka Tilla | 2.28 | – | Timarpur | 2.73 | – |
| Shalimar Village | 3.69 | – | Naharpur | 7.46 | 108 |
| Rohini Sec-2 | 2.4 | – | Maharani Bagh Market | 3.99 | – |
| Zakhira | 2.63 | – | Rana Pratap Nagar | 2.28 | – |
| Inderpuri | 1.67 | – | Lakshmi Bai Nagar | 3.65 | – |
| Pragati Vihar | 6.43 | – | Jeevan Nagar | 3.23 | 232 |
| Panchkuian Road | 2.38 | – | Kishan Ganj | – | 190 |
| Kotla Mubarakpur | – | 328 | Sanjay Camp | – | 240 |
| Sudershan Puri | – | 202 | Shantivan | – | 144 |
| Okhala Ind. Phase-1 | – | 110 | Sarita Vihar | – | 149 |
| New Friends Colony | – | 102 | Naraina | – | 600 |
| Begampur | – | 240 | Moti Bagh Part-1 | 2.11 | – |
| | | **Block-Shahdara** | | | |
| Ghazipur | – | 139 | E. Vinod Vihar | – | 180 |
| Durgapuri Chowk | – | 141 | | | |
| | | **Block-Mehrauli** | | | |
| Dera Gaon | – | 690 | Rajokari Road | – | 262 |
| Gadaipur | – | 743 | Jaunapur | – | 124 |

*Source*: Central Ground Water Authority (1998), New Delhi. Maximum Permissible Limit of Fluoride: 1.5 mg/1; Nitrate: 100 mg/l.

## Groundwater Management

Periodic assessments and measurements of groundwater level provide accurate situation and changes in the quality taking into consideration the economic viability. Utilization of groundwater resources should be so regulated so that as not to exceed the recharging possibility. Groundwater recharge projects should be initiated for augmenting the available supply. Integrated development of surface water and groundwater and their rational use should be encouraged right from the project planning stage and should form an essential part of water resource development programme. A phased programme should be implemented for improvements in water quality. Urban development and related activities should be planned with due regard to the constraints imposed by the configuration of water availability. There is a need for water zoning in metropolitan region and the economic activities should be regulated in accordance with such zoning (Government of India, 1987). This is supplemented by waste management strategy. Groundwater resources may be planned to meet the mega city needs for the next fifty years. Serious consideration is needed to be given for storage of rainwater in low lying areas in Delhi and surroundings. Protection of groundwater is essential. Aquifers may be recharged by harnessing rainwater. Recharging aquifers with the effluents could also be experienced in Delhi. Environmental improvement along roads and majors drains is needed (Asian Development Bank, 1993). Water quality study using remote sensing technique is still in its infant stage and lots of research and development has to take place before converting into an operational activity.

## REFERENCES

Asian Development Bank (1993): Managing Water Resources to Meet Mega city Needs Case Study Delhi, *Proceeding of the Regional Consultation*, Manila, 24-27 August.

Central Board for the Prevention and Control of Water Pollution (1985): Groundwater Quality in the Union Territory of Delhi Abridged Report and Programme objectives series, New Delhi.

Central Ground Water Board (1998): Report on Ground Water Resources, its Depletion and Contamination in Najafgarh Block of NCT Delhi, New Delhi.

Central Pollution Control Board (1998): Report on Ground Water Pollution in NCT of Delhi, New Delhi.

Central Pollution Control Board (1995): Status of Water Supply Wastewater Generation, Collection, Treatment and Disposal in Metro cities (1994-95), Control of Urban pollution series CUPS/42, New Delhi, 1-38.

Central Pollution Control Board (1979): Industrial Survey, UT of Delhi Control of Urban Pollution Series, New Delhi, 1-10.

Government of India (1987): National Water Policy, Ministry of Water Resources, New Delhi.

Hilding-Rydevik, Tuija and Johanson, Irene, (eds.) (1998): *How to Cope with Degrading Groundwater Quality in Europe*, Swedish Council for Planning and Coordination of Research, Stockholm.

# 12

# Impact of Metropolisation on Land Use and Socio-economic Conditions in Urban Fringe of Delhi

## *Case Studies*

Anju Singh

### Introduction

The villages around Delhi are changing in terms of their shape, size and occupational characteristics owing to the city's overwhelming impact in the form of its dynamics of expansion. The other factors affecting these villages are the acquisition policy of the development authorities on agricultural lands, which has brought change in the landscape of the villages. The relatively independent settlements on the fringe of urban limits have been transformed into dependent isolated developments, not properly integrated with urban limits. These villages still play an important role in supporting the city of Delhi. They provide low cost accommodation to the incoming migrants and a variety of services to the more affluent of the city. It is a process that has morphological, social and economic aspects. In the Indian context Ramachandran (1991) identifies five stages in the transformation process: (1) the rural stage, (2) the stage of agricultural land use change, (3) the stage of occupational change, (4) the stage of urban land use growth, and (5) the urban village stage. Consequently, a stage-model should be interpreted carefully as stages might overlap both in space and time in some villages, while others go through all the stages. Therefore, we prefer not to use the concept of different stages, but of processes of agricultural land use change, occupational change and urban landuse growth in the rural-urban fringes of big cities (Druijven and Singh, 1996, 1998; Potter and Salav, 1990).

## Development Pressures in Delhi Metropolitan Region

In Delhi National Capital Territory, village land is of two categories: '*abadi*' area (village settlement) and agricultural land. The '*abadi*' was marked for village settlements and included land for houses and common needs of the people; it was marked in red and was known as the '*Lal Dora*'. This '*Lal Dora*' is the settlement area as defined in Master Plan of Delhi, 1980-81. The limits of the settlement area were determined at the time of settlement operation. Regular survey maps of settlements were made in the year 1864 and subsequently in 1880 and 1908-09. The settlement area was allotted field numbers. But after these maps of settlement area were made, the area was not reviewed again.

Under the provision of the Delhi Land Reform Act, 1954, the land, outside the '*Lal Dora*' for purposes other than specified was not prohibited but also penalized. This land was prone to high densities within the settlements. In 1954 the Consolidation of Holdings (Prevention of Fragmentation) Act was passed. This act provided the extension of the existing settlements, with the consent of the villagers. But this act only provided for the land-owners desirous of owning plots for residential purpose in the extended settlement limit subject to their surrendering agricultural land (2 or 3 times area of required plot). Thus, this act never capitalized.

The villages, which fall within the urbanised limits of the city area cease to be rural area by virtue of the notification, issued under Section 507(A) of the MCD Act, and their rural status, including '*Lal Dora*' exemptions become redundant. However, the strong nexus of officials, property dealers and politicians prefer to perpetuate this conception with regard to definition of urban village and cognitive skinning these settlements for profit.

Demographically, the city's population stood at 0.21 million in 1901 and rose slowly to a figure of 0.70 million by 1941. But after partition of India, there was a sudden influx of refugee migrants. As such, during a brief period of only four years after partition, the population of the city more than doubled itself to reach the figure of 1.44 million in 1951 (Delhi Master Plan, 1962). This scenario alarmed the planners. The sudden spurt in Delhi's population growth rate and haphazard urban sprawl forced the Government of India to take steps urgently for planned development of the city. The Delhi Metropolitan area covering the area up to approximately 20 kms. around the developed urban limits of Delhi (about 2000 km$^2$) was segregated as the immediate influence area. This segregation should be based on the criterion of economic interdependence between Delhi and the town in the region and the intervening agricultural areas. It was envisaged that in the wider region (National Capital Territory of Delhi), the selected urban centres would be identified for induced growth to help in deflecting population that would otherwise be attracted towards Delhi. Thus, a regional approach was adopted for controlling the city's population and chalking out a strategy for its planned growth.

Over the last two decades there have been some of the fastest growing areas in Delhi in terms of both population and employment. These growth areas lie beyond the '*Lal Dora*', which were established in Delhi during the 1971 and 1981 to give breathing space to cities and to halt the urban sprawl of the 1947. Therefore, new housing and centres of employment were prevented in the natural land, i.e. Delhi ridge forested areas which were to act mainly as the 'lungs of Delhi'. In some ways, it indicates that little development has taken place within them. While in other ways it represents their failure to halt the spread of the metropolis in functional terms, since people now commute across the '*Lal Dora*' to the central city. Given this functional extension of the cities beyond their '*Lal Dora*', it is not surprising that development has exerted great pressure on the DDA and NCR planning authority to allow them to create houses and jobs within and on the edge of the '*Lal Dora*'. They see them as areas of high growth potential, particularly the Delhi Metropolitan Ring. In brief, the conflict is whether the Delhi ridge remains green or became gray. Present Delhi metropolitan ring extending over about 2,000 sq. kms. was delineated as the immediate influence area. This extending trend of Delhi metropolitan ring is grabbing the Rural Delhi areas. During two decades 1971-81 and 1981-91 area of Rural Delhi has been continuously decreasing at a raster rate like 14.24 per cent and 13.24 per cent, respectively.

## Impact of Metropolisation on Surveyed Villages

For understanding the impact of metropolisation on urban fringe area, the fringe villages have been divided into two categories, i.e. rural dominant and urban dominant fringe villages. For primary survey 3 villages (150 households, 50 from each village) from Alipur Block has been selected for first category. The village Chandan Hola (50 households) of Mehrauli block has been selected for second category (Fig. 1).

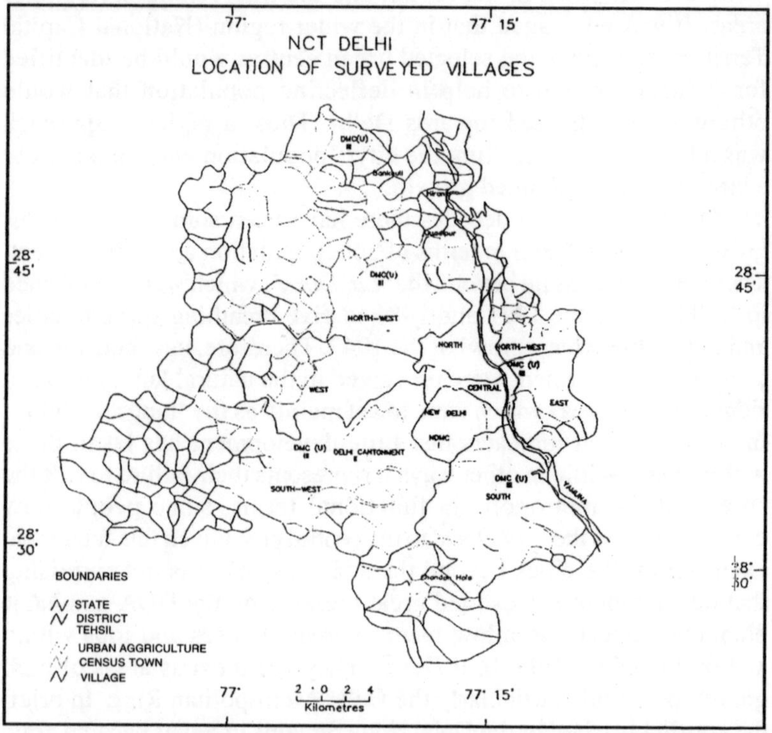


**Fig. 1:** Metropolisation on surveyed villages

## Case of Agriculture Dominant Villages

Presently, Delhi NCT is an amalgamation of large number of traditional settlements that include rural, urban and other typology which is in combination of both urban and rural is popularly known as "urban village". The term "urban village" was first coined in the master plan of Delhi (1962) for the villages in urban fringe where rural types of industries together with the population engaged in such activities exist together. This had been similar to the European concept of urban villages round the metropolitan center, where rural kind of activities had been planned to serve the city (Hardy *et al.*, 1992; Barrow, 1991).

There are very few villages in Delhi's fringe, which are purely rural in character. But what is essentially missing from a few villages is the daily interaction with the city. Lack of proper bus service is one of the main reasons for the lesser development and neglect of the village from urban forces. Two villages are in the stage of occupational change Hiranki and Bankauli. One village, Qadipur, is in the stage of urban land use growth. These three villages have been selected for case study.

The metropolitan fringe exhibits a high density of occupied dwellings. The medium density of the rural-urban fringe is of high proportion in residential, commercial, industrial areas, which are distinct from farmland with a higher rate of increase in population density coinciding with land use conversion and commuting. The metropolitan fringe characteristics are a result of institutional factors. For convenience of vast space requirements, many of the public utilities serving the cities are being located in the metropolitan fringe areas, like recreational land, water supply stations, radio stations, sewage disposal plants, airports and cemeteries. Other necessary but unwanted uses are slaughter houses, wholesale markets, oil storage, noxious industries of all sorts, junkyards, substandard dwellings, residential colonies and the brick kilns etc. Such activities escape urban control as located in the nearby countryside where rural land use stops near the outskirts of city land. Since, urban land planning is confined to city limits only, this rural-urban fringe is neglected in this respect (Notan and White, 1984).

These transformations are representing all undesirables but essential land uses, relevant for the normal life of urban areas. These are pushed towards the city's outer limit and often into the fringe villages. The classic examples of such land use are garbage dumps, water works, sewage disposal tanks and forms, burial and cremation grounds, airports, timber-yards, brick kilns, etc. The fringe area is considered suitable for relocation of city slums. The impacts of these processes are felt more in the urban fringe areas. The rural-urban fringe is an identifiable zone, distinguishable from both city and countryside by its heterogeneous and dynamic character. This heterogeneity is seen in land use, economic activities, modes of production and social and political interactions, etc. In addition to this heterogeneity, the fringe is also characterized by its dynamism.

The urban fringe is therefore a zone of transition, which is in a permanent state of influx in both space and time (Blaikie and Brookefield, 1987).

When the traditional village settlements were urbanised, they were altered from their original rural character. Their function changed from producers to that of consumers as their agricultural land was taken to fulfil the needs of the growing city. This transition was taking place due to natural responses to urban pressures. The government, while implementing the schemes for development, very often created wrong pressure at the wrong time. This resulted in a degraded village environment. Keeping this process of transformation of villages, Alipur block has been selected for the study. The block is located in the metropolitan fringe of Delhi NCT. Geographically, Alipur is located in Northern part of Delhi in between the Narela ring town and Delhi Metropolitan City.

The block consists of many villages but this study has taken three villages from the block, i.e. Bankauli, Qadipur, and Hiranki. These sample villages have been selected on the basis of the following criteria: (1) distance from the city, (2) number and sizes of households of the village, (3) land use changes. Bankauli village is situated closer to the city. The rest two villages lie in the peripheral area of Delhi Metropolitan City. The expansion of city limits on agricultural land is a pressing problem in the region. This describes a detailed scenario of changing land use pattern for three-selected sample villages of Alipur Block. These villages are most affected by the process of rural-urban fringe expansion, due to its closer proximity to Delhi metropolitan region. These areas consist of settlements, which predominantly have rural character, but are becoming more and more integrated into the city. This is characterized by the changing land use. Land use is changing primarily from agricultural to industrial plots, warehouses, rice mills, residential colonies etc. villages like Hiranki, Bankauli, Qadipur have been most affected due to the extension of urban areas into the village land.

The following parameters emerged from the rural households:

1. Presence of varying educational achievements (mostly low);
2. Difference in age, education, knowledge, experience, cultural background, caste, incomes, etc.;
3. Often conservative in their values and practices;
4. Have a wide range of farm and home situation;

5. Influenced by many different groups in this increasingly complex society;
6. The levels of asset ownership often dictated the level of living and education;
7. A largely male dominated society with less preference for female literacy as well as economic independence.

***Land Use Systems and Patterns***

Land use for residential development or other urban purposes are much more profitable than land use for agriculture. The prices paid to farmers for urban land use or "other non-farm purposes ranged from about four to over hundred times than those paid for agricultural purposes. With such disparities in urban land use it will almost invariably displace agriculture. The mere anticipation of urban development on an area of land increases its value. If this expectation is strong, urban developers or speculators may purchase the land from the farmer, and be content to leave the land unused until it is developed, reaping capital appreciation rather than steady income. If the expectation is rather more vague, the land may not change hands, but the farmers may be reluctant to invest capital and effort in maintaining levels of fences fertility, drainage or other capital equipment, and in particular farmer may be reluctant to invest in land such as orchards which requires a number of years to mature. Agricultural land use may therefore, be discontinued, modified, or disintensified ahead of actual urban encroachment. The inner most ring of rural land, therefore, may not be given over to highly intensive agriculture or to the production of poorly transportable commodities, but it may instead be characterized by a lower level of agriculture activity than that prevailing further from the advancing city.

The process of planning and land use allocation is too slow, which may result in contradiction between planning proposals and reality. Planning for urban fringe may involve long-term as well as short-term strategies. Long-term strategies govern the consideration for overall structure of the city, where as short-term strategy would be to make land available for immediate need of the city. The rural-urban fringe is an area that is constantly facing pressures by the growth of the city and is actually a part of the growth process. Since 1900, the urban area of Delhi has grown nearly many folds, thereby, absorbing more than 100 villages (Hardy *et al.*, 1992). The urban area of Delhi has increased by 33 per cent at the cost of the rural area during the period of 1971-91.

### *Social Structure and Land Ownership*

Brahmins are traditionally not the land cultivating castes. This fact is reflected in this table where only 58 per cent of Brahmins own the land. Even among these, most of their land is less than two acres of land and most of which is gifted to them for their services towards religion. Mostly Jats are the major land-owners. More than 95 per cent of the Jats own land. Their amount of land-ownership in all the land classes through 0-1 acres land is the most widespread. This refers to the subsistence level of farming present in the villages. Rajputs are the more prosperous land-owners. Most of them own landholdings, which are between 2-3 acres and some of them, own more than 4 acres. Banias are very few in numbers and they own land less than 2 acres. As they are the trading castes, land is of little value to them. So they lease out land either on rent or use it for subsistence. OBCs, which includes the Yadavas, are traditionally the cultivating caste who had turned up to non-agricultural activities. Most of the SCs own that land which was allotted to them during the 20-Point Programme (Table 1).

Almost 70 per cent of the residents of Alipur block owned land and the chief among them are the Jats who form 40 per cent of the land-owners. They own the land, most of which is more than 4 acres in size. A large percentage of SCs and OBCs do not own land.

**Table 1:** Alipur Block: Social Structure of Land-ownership.

*(In Percentage)*

| *Castes* | *Not Own Land* | *Own Land* | *Amount of Land-owned by Land-owners* | | | | |
|---|---|---|---|---|---|---|---|
| | | | *0-1 acre* | *1-2 acre* | *2-3 acre* | *3-4 acre* | *> 4 acre* |
| Brahmin | 42 | 58 | 15 | 30 | 36 | 4 | 15 |
| Jat | 4 | 96 | 5 | 30 | 25 | 12 | 28 |
| Rajput | 9 | 91 | 22 | 16 | 34 | 3 | 25 |
| Bania | 12 | 88 | – | 86 | – | – | 14 |
| OBC | 73 | 27 | – | 50 | – | – | 50 |
| SC | 62 | 38 | 54 | 38 | 8 | – | – |
| Total | 30 | 70 | 15 | 31 | 26 | 6 | 22 |

*Source*: Primary survey.

***Land Transfer***

As apparent from land-ownership, Jats are the most dominant land-owners. They are also the major land sellers. A total of 58 per cent of land-sellers are Jats. Nearly 8 per cent of land-sellers are SCs who have sold out their land, which was less than 4 acres. Maximum sale has taken place in the 1-2 acre category, which refers to the small and marginal farmers who are selling off their lands because of water scarcity and salinity (Table 2).

**Table 2:** Alipur Block Social Structure and Land Sale.

(*In Percentage*)

| *Castes* | *Percentage of Social Structure of Families Selling Land* | *Amount of Land Sold by Land-owners* | | | | |
|---|---|---|---|---|---|---|
| | | *0-1acre* | *1-2 acre* | *2-3 acre* | *3-4 acre* | *>4 acre* |
| Brahmin | 18 | 12 | 58 | 12 | 6 | 12 |
| Jat | 42 | 3 | 18 | 15 | 36 | 28 |
| Rajput | 25 | – | 4 | 48 | 4 | 44 |
| Bania | 5 | – | 100 | – | – | – |
| OBC | 2 | 50 | – | – | | 50 |
| SC | 8 | 83 | – | – | – | 17 |
| Total | 100 | 10 | 25 | 25 | 9 | 31 |

*Source*: Primary survey.

In Alipur, Jats own maximum land and have even contributed most to the total land sales in the block. Maximum sales are for land holding 3-4 acres in size while, Rajputs have sold maximum in the 2-3 acres plots. Maximum sales have taken place in plots more than 4 acres in size. In one village 2 Rajput families had together sold 45 acres, contributing to the creation of a separate class of 20 acres. Most of the SCs have sold off their allotted land.

***Fragmentation of Land-holdings***

The land is fragmented among the members of the same family. It is evident that maximum numbers of plots fall under the first category followed by second, third, fourth and fifth, respectively. It represents that the family size increases and the ancestral land gets divided among the family members thereby leading to the fragmentation of the land.

Consequently, land use patterns are changing. The larger numbers of small plots are seen in Bankauli village (68%) followed by Qadipur (67%), Hiranki (54%). Bigger sizes of plots are observed in Qadipur village (6%) followed by Hiranki (4%) (Table 3). It is an area of opportunities; but wasted opportunities. It is an area of resources; but unmanaged resources with increasing population growth and material needs of modern living, which will increase the demand of land for housing and industrial space.

**Table 3:** Fragmentation of Land Holdings.

(*In Percentage*)

| *Villages* | *0-1 acre* | *1-2 acre* | *2-3 acre* | *3-4 acre* | *> 4 acre* |
|---|---|---|---|---|---|
| Hiranki | 54 | 32 | 10 | – | 4 |
| Bankauli | 68 | 32 | – | – | – |
| Qadipur | 07 | 7 | 7 | 13 | 6 |

*Source*: Primary survey.

## *Sale of Land*

The domicile of the buyers play an important role in the probable use to which the land shall be put to. A local resident is most likely to continue with agriculture but may be in future utilize it for other uses. An outsider is likely to use the land for non-agricultural purposes and even pays more for it.

In all three surveyed villages in Alipur Block, the maximum amounts of land are sold to outsiders and very few amount of land has been sold to local people because outsiders pays more for it. The high percentage shares are found in Bankauli (100%) followed by Hiranki (95%) and Qadipur (80%) (Table 4). This is so because many cities are surrounded by productive farmland, therefore, when cities expand they frequently take over good quantity of farmland.

**Table 4:** Sale of Land.

(*In Percentage*)

| *Villages* | *Local* | *Outsiders* |
|---|---|---|
| Hiranki | 5 | 95 |
| Bankauli | – | 100 |
| Qadipur | 20 | 80 |

*Source*: Primary survey.

### *Present Use of Sold Land*

The maximum amount of sold land has been presently utilized for farmhouses, residential colony and commercial/industrial activities. Mainly farmhouses in Hiranki among the three selected sample villages have utilized the land sold for non-agricultural activities. In Hiranki, the percentage share was 60 per cent followed by Bankauli (56%), and Qadipur (14%). Land utilized for colony purpose of which the maximum been observed in Qadipur 57 per cent followed by Hiranki (32%) and Bankauli (3%). In Bankauli 25 per cent and Hiranki 4 per cent of sold land have been utilized for commercial/industrial purposes (Table 5). Largest of the uses of sold land has been for farmhouses, warehouses, etc. These are signs of suburban development. Good quality agricultural land fetches better prices when the ultimate use is for farmhouses.

**Table 5:** Present use of Sold Land.

*(In Percentage)*

| *Villages* | *Farm-house* | *Colony* | *Agriculture* | *Commercial & Industrial* | *Others* | *Nursery* |
|---|---|---|---|---|---|---|
| Hiranki | 60 | 32 | – | 4 | – | 4 |
| Bankauli | 56 | 3 | 16 | 25 | – | – |
| Qadipur | 14 | 57 | – | – | 29 | – |

*Source*: Primary survey.

### *Mode of Access to Water*

A number of municipal connections, hand pumps, water tankers, wells, tubewells, public water taps and other sources are available in Delhi fringe area (Misra, Singh and Malik, 1998). Bankauli village, with a high concentration of well-off households benefits on an average from a better water supply. These villages enjoy maximum water from municipal connection. It has been observed that Bankauli has municipal water supply of 77 per cent among the all other sources and is followed by Hiranki (43%), Qadipur (33%) respectively. The surveyed villages have received water from all these sources. But the mode of access to water varies from village to village. In Qadipur village, most of the households receive water from water tanker, because in this village most of the people are residing in *jhuggi jhopri*. (Table 6).

**Table 6:** Source of Water.

(*In Percentage*)

| *Villages* | *Qadipur* | *Hiranki* | *Bankauli* |
|---|---|---|---|
| Municipal supply line | 33 | 43 | 77 |
| Water Tanker | 31 | 9 | 6 |
| Hand Pump | 33 | 25 | 10 |
| Tubewell | 2 | 22 | 2 |
| Public Water Tap | 2 | 2 | 2 |
| Others | – | – | 2 |

*Source*: Primary survey.

The main source of urban water supply is from the Government than from private and other sources. These surveyed households are using water from government source, i.e. MCD water supply. Out of three village sources Hiranki received higher water supply (32%) followed by Qadipur (30%), and Bankauli (10%) from private sources. The other water supply agencies are also playing very important role but the working pace is slow (Table 7).

**Table 7:** Source of Water Supply.

(*In Percentage*)

| *Villages* | *Qadipur* | *Hiranki* | *Bankauli* |
|---|---|---|---|
| Government | 52 | 49 | 88 |
| Private | 30 | 32 | 10 |
| Owned by others | 10 | 19 | 2 |
| Illegal | 4 | – | – |
| NGO provided | 4 | – | – |

*Source*: Primary survey.

## Impact of Metropolisation on Urbanized Village: A Case of Chandan Hola Village

The Chandan Hola village is situated at the fringe of Delhi and has experienced the impacts of metropolisation in various ways. During the survey in total 50 households were selected on the basis of random sampling method. Muslims dominate the village, about 88 per cent households are Muslims, and the rest 12 per cent are Gujars. Illiteracy is still high in the village, 48 per cent are illiterate and only 4 per cent respondents are educated upto higher secondary level (Table 8). There has not been much impact of metropolisation in terms of families' pattern in this village. Joint families out number nuclear families.

**Table 8:** Sources of Income.

| *Source* | *Percentage of Respondents* |
|---|---|
| Agriculture | 28 |
| Service | 16 |
| Business | 36 |
| Agriculture + Business | 4 |
| Agriculture + Service | 4 |
| Others | 12 |

*Source*: Primary survey.

### *Occupation and Income Structure*

The trend of urbanisation and metropolisation has clear-cut impact on the occupation pattern of this village. Agricultural activity has become secondary. Business, especially the sale of milk and milk products to the nearby markets, has become the primary activity of the families. People are switching over to dairy farming as it provides them cash income. Service, both in government and private sectors, is at the third place. Most of the people in this village belong to lower income group having a monthly income in between Rs. 2000-5000. Only 4 per cent of the respondent earn above Rs. 10,000 per month (Table 9).

**Table 9:** Income Structure.

| *Income (in Rs.)* | *Percentage of Respondents* |
|---|---|
| Below 2000 | 8 |
| 2000-5000 | 64 |
| 5000-10000 | 24 |
| Above 10000 | 4 |

*Source*: Primary survey.

### *Land Use Dynamics*

The land use pattern of this village is changing very fast due to metropolisation. Agriculture is being replaced by dairy and allied activities. Only 28 per cent land is used for agricultural activity and that too for the subsistence. People grow wheat, Jawar, Bajra and Mustard for their own use only. House, building, market complex

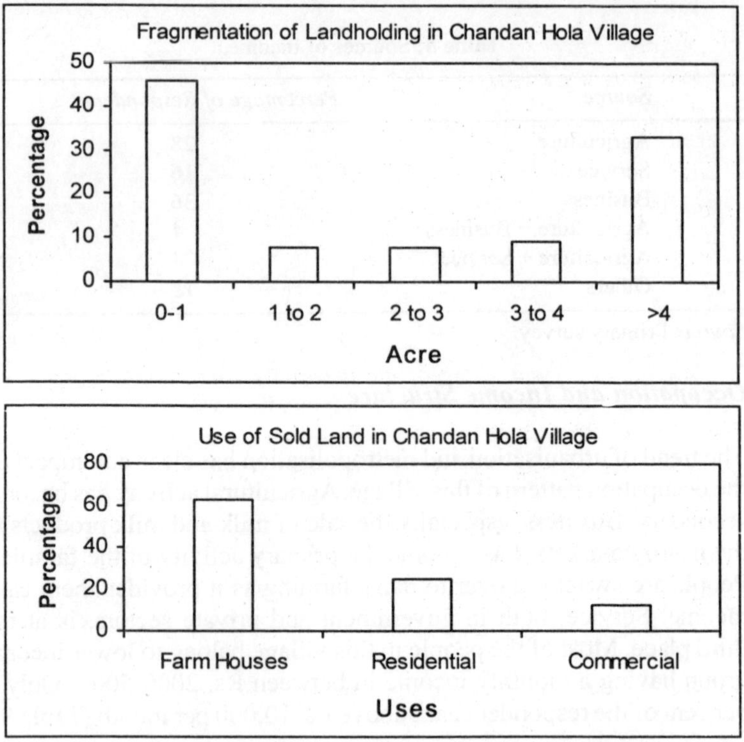


**Fig. 2:** Fragmentation of Land Holding in Chandan Hola Village

and shops constructions are the main activities and most of the land is being devoted to such activities. Most of the agricultural lands have been sold out and new buildings/houses are being constructed on those lands (Fig. 2). Some agricultural lands have also been converted into farmhouses. Size of the existing landholdings is also very small. Mostly the size of the landholding is below one acre. Medium size of land holdings have been generally sold out (50 per cent of the total sold land comprises of this medium sized land holdings of 2-3 acres), remaining are used mainly for kitchen gardening. Most of these landholdings were sold before 25-30 years, the period which, may be considered as the beginning of real out expansion of city boundary. One interesting fact of this land sale and purchase is that most of the purchasers of these lands are outsiders. It has also been found during the field survey that many of the farmers of these region sold their land at higher prices and bought more amount of agriculturally productive lands in nearby State of Haryana.

Another important feature of this village is that still fuelwoods are in vogue and these are putting serious threats to the surrounding Ridge. The people in this village are not very much aware about the health and educational facilities. They are still practicing the traditional village culture.

### *Water Facilities, Sanitation and other Household Facilities*

All the houses in this village are pucca (cemented) Proximity to the city and impact of urbanisation is reflected in the housing quality. All the houses have proper sanitation facilities, like toilets, etc. The village has also good water supply. 80 per cent households have access to tap water while only 20 per cent have tubewells and handpumps. People in this village are satisfied with water condition. But they are very much concerned about the deteriorating water table. Excessive exploitation of water is the main culprit of deteriorating water table (53% respondents), while 48 per cent claim shortfall in precipitation is the main factor of deteriorating water table. Although, agriculture is not widely practiced, but whatever land is given to agriculture is generally irrigated by tubewells. Waterlogging is the pregnant problem in the village. Although, there is no concern about the waterlogging and health risks, but this can be the major potential area for health hazard. Every house in this village is electrified. But people in this village use different means for cooking purposes. This aspect does not reflect impact of urbanisation and metropolisation. Urban people use LPG for cooking purposes but people in this village are still using wood and cowdung cake.

## Metropolisation, Environment Degradation and Urban Fringe

Due to the spread of urbanisation in Delhi, the urban areas are expanding in its physical aspect and it is slowly encroaching upon the rural agricultural lands. These villages which come under the rural-urban fringe of Delhi are thus experiencing rapid change in the distribution of land for different purposes. It has been observed during the survey that there is a common notion among the people of these villages that their land will be sold some day or the other and they will get good remuneration out of that land. This can be a possible reason behind the extensive use of agricultural land by greater use of technological innovations like fertilizers, pesticides, multi-cropping (growing more than one crop a year) without giving a thought to its proper usage for a sustainable agricultural development. This results into the exhaustion of soil fertility and loss of soil nutrients. Alkalinity and salinity is the result of the unscientific irrigation methods. When a farmer finds that the income from the agricultural land is not remunerative, he sells off his land to outsiders who give the highest bid for his land. These dealers either sell off their land again to the highest bidder or he builds residential colonies on the land. Thus concretisation takes place and it results in overcrowding of the area as a result of urbanisation. Various types of pollution among which land degradation and air pollution are seen as the most important and fastest growing threats follow urbanisation. Since there is no adequate perennial source of water to feed the population of the area, underground water is the only source left to bank upon, which results in the over-withdrawal of fresh groundwater without being recharged. These have resulted into the water table going down and increase of saline water table.

The second important problem is the disposal of the sewage from village drains. With the increase in the village population, the amount of domestic and local industry's effluents have also increased. With the absence of sewers in most of the villages, the domestic and industrial waste is channeled into the local water bodies known as the "*jhors*". These water bodies, which earlier were the lifelines of the village and served as the bathing ponds of both the cattle as well as local inhabitants, have. now become nothing more than dirty pools of stagnant water emitting poisonous and fatal fumes and gases.

The closer proximity to Delhi metropolitan ring and Narela causes land degradation due to constantly changing pattern of land occupancy and land use. Various land uses have evolved over the agricultural land, e.g. ware houses, construction material industry like the brick kilns, and manufacturing industries. All these have a bearing on the quality of agricultural land.

The problem of renewing land for agriculture ensures that it remains economically viable. Agriculture is a poor competitor in the category of remunerative land uses in the metropolitan fringe. The farmers have reduced the amount of inputs used for the land development because farmlands are small and crop production has become very intensive as well as expensive. This causes decline in fertility in the long-term. Hence, there is a fall in productivity and generally there is reduction in outputs from the farming areas. Moreover, delay in development causes fallow or idle land or abandoned farms. Expansion of Delhi metropolitan ring may absorb parts of farms. The whole farm therefore generates less. There is tendency to farm decays or land degradation.

## Conclusion

The rapid growth of land use changes occurs in Qadipur and Bankauli. These villages are located on the main route of Delhi metropolitan ring and in closer proximity to the metropolitan area. Great difficulties arise in the metropolitan fringe due to the intermixture of non-conforming land uses with other rural uses.

Non-conforming activities in fringe areas occur for purely economic reasons because space needs are high and their competitive prices are low in Delhi metropolitan fringe area. For example, water storage service and Government depots, large manufacturing plants, public utility installations, heavy transport parts, cheaper land are beyond the city outskirts. These non-conforming activities create conflicts and pressures when put together with residential land in rural fringe areas.

There seems to be no direct and apparent relationship between water quality and its socio-economic implications on society. A further probe and data collection coupled with extensive field observation shows that water quality and availability plays an important role in modifying the attitudes, awareness and means of livelihood in the rural-urban fringe. As is apparent through the data, there is a direct relationship between agricultural development and its importance as a means of livelihood and good water quality and easy availability. However, the importance of agriculture as a means of livelihood declines in the areas where water is scarce and of poor quality.

The lack of public insight and absence of serious concern about the magnitude of the looming fresh water challenges is devastating, especially in the fringe areas of Delhi and all other metropolitan cities. All over, the world an understanding of this valuable resource is taking place. The study indicates an existing will to act but are still failing to implement sustainable solutions. A major initiative is called for to bring about the needed changes and to mobilize public opinion: a freshwater revolution.

The analysis raises several questions to be examined, amongst which, one of the most important is to find out whether it is possible to provide a regular water supply in fringe villages. The municipal authorities emphasize the lack of water resource as their main constraint. Yet, a poor management of the distribution network mostly causes inadequate supply. One can also presume that a continuous supply would result in a reduction of the costs borne by households. Thus, a rationalisation of the distribution network could be a first step towards service improvement if it is financed partly by a transfer of the costs of unreliability.

Improving the quality of supply is essential in order to regain the user's confidence and to create favourable condition, for a transformation of the water supply services. It can also play a role in the conservation of groundwater table at a time when its depletion is accelerating and threatening the sustainability of the city's economic development. A better quality of service, which would reduce individual usage of wells is, however, not enough legislative changes on rights to groundwater are necessary. They are envisaged but they should be part of a larger effort on changes, of institutional structures and on the possibility of better coordinated infrastructure policies.

This we see that the Delhi Metropolitanism in its rural-urban fringe has revealed map and hyperactivity of land-dynamics and the water-available in this area acts as a catalytic factor in its sustainability.

## REFERENCES

Barrow, C.J., (1991): *Land Degradation Development and Breakdown of Terrestrial Environment,* Cambridge University Press, Cambridge.

Blaikie, P. and Brookefield, H. (eds.), (1987): *Land Degradation and Society*, Methuen Publication, London.

Bryant, C.R., Huigen, P.P.P. (eds.). (1998): *Dimension of Sustainable Rural Systems,* Nederlandse Geografische Studies 244, pp. 105-114.

Bryan, J. (ed.) *Impacts of Urban Growth on Surface Water and Groundwater Quality*, IAHS Publication, No. 259.

Delhi Development Authority (DDA), (1962): *Master Plan for Delhi,* New Delhi.

Druijven, P.C.J. and Singh, R.B., (1996): Environmental Degradation and Its Impact on Livelihood Strategies in The Urban Fringe of Delhi, In: R.B.

Singli, (ed.), *Disasters, Environment and Development,* New Delhi: Oxford and IBH Pub.

Druijven, P.C.J. and Singh, R.B., (1998): The Depletion of Common Lands in The Rural-Urban Fringes of Delhi: A Case Study of Unsustainable Development in Bowler, I.R.

Hardy, J.E., Mitlin, D. and Satterwaite, D., (1992): *Environmental Problems in the Third World Cities,* Earthscan, London.

Misra, B., Singh, R.B. and Malik, Anupma, (1998): Delhi: Housing and Quality of Life in Misra, R.P. and Misra, K. (eds.). *Million Cities of India,* Sustainable Development Foundation, New Delhi.

Notan, P. and White, G., (1984): Urban Bias, Rural Bias or State Bias: Urban-Rural Relations in Post-Revolutionary China, In Harris, J. and Moore, M. (eds.) *Development and the Rural-Urban Divide,* Frank Cass, London, pp. 52-81.

Potter, R.B. and Salau, A.T., (1990): *Cities and development in the Third World,* Mansell, London.

Ramachandran, R., (1991): *Urbanisation and Urban System in India,* Oxford University Press, New Delhi.

Singh, R.B., (1999): *Urban Impacts on Groundwater Quality in the Delhi Region* In Ellis.

# 13

# Urban Growth and Land Use Problems of a Hill Town

## *A Case Study of Darjiling*

SUSHMA BOMJAN

### Introduction

Man is an active agent who influences all the natural systems and also the main user of natural resources. Therefore, positive or negative impacts of human activities are dependent upon man-environment adjustment. Any developmental planning is directly related to and dependent on its people. To analyse the problems of urban growth and urban land use of any hilly towns, it is important to know the demographic structure and dimensions of economic development of these regions.

In the study of urban land use of Darjiling Himalayan region, discussion on demographic trends becomes an important feature to understand; their impacts on society, economy and political set up at its present context. The details of demographic trends are useful in assessment of the impact of population growth on land use directly or indirectly. Population growth is a powerful force to bring about a change in traditional methods and to transform the economy into rather advanced and productive stages.[1] Thus, the study of demographic structure and the socio-economic characteristics of the hilly towns is essential in all developmental planning.

The Darjiling Himalayas, a segment of the Eastern Himalayas, is a famous tourist hill-resort established by Britishers in 1835. The then Superintendent Dr. Campbell had reported the population of

Darjiling to be 1900 in 1850 and 2200 in 1869. Within short span of time this region has witnessed tremendous growth of population stemming mainly from immigration which is 16 lakhs in 2001 census. Majestic towering snow-capped peaks, scenic grandeur, torrent streams, bracing climate, varied flora and fauna—all have contributed to its rapid growth and development. Tremendous population concentration, within a short span of few decades, have created manifold problems—political, economic, social, cultural and ecological disharmonies in this region. Accelerated population pressure on limited available land has led to indiscriminate exploitation and ruthless destruction of natural resources. This has led to serious ecological damages and far reaching consequences, visible through bare hill slopes, stony and gravelly riverbeds, erratic river courses, change in climatic and hydrological regimes, extensive landslips, disappearance of rich flora and fauna etc. Man, the worst enemy, has taken a heavy toll of the natural environment all around and faces the worst nemesis due to landslides.

The socio-economic system of a place is certainly influenced by many factors among which the historical growth of the place needs to be examined first, includes physical and economic growth and cultural set up.

The population and urbanisation is a major constitution of the socio-economic system. Among these, population growth occupies important seat. Darjiling was totally an uninhabited forest area until the Britishers acquired this territory in 1835. About 160 years ago Darjiling was not as populated as it is now and obviously the demographic pattern has also undergone changes. To understand the changed pattern of a town one has to understand the causes of the urbanisation process and growth of urbanisation as a whole.

Urbanisation is a process by itself bringing changes in the society and this is clearly evident in the life and activities of the people inhabiting the Darjiling Himalayas. As a process, here, urbanisation has not been the result of a surplus economy as is usually the case in any sedentary community, instead, it has been created in different manner if we look into the background (history of its establishment). It all started with the British idea of building a sanatorium. This place made a strong impression on the Britishers as the most suitable place for health resort, eventually turning out to be a famous hill resort of India. Apart from thinking of as a sanatorium and a health resort, it was the strategic location of this place, occupying significance. This place provided direct access to the hills and opened up enormous avenues in both economic and strategic development of the region (Fig. 1).

## Growth of Population

The entire Darjiling area was under forest. "It was estimated that the whole track comprising 357 km$^2$ contained only 100 souls."[2] There were around 100 Lepchas around the present Darjiling town. The area was sparsely populated and inhabited by Lepchas and Bhotias. With the exploration of the area and with the widespread developmental activities the inhabitants eventually increased. The growth of population was very high between 1872 and 1991 (Table 1). Again, relatively higher growth can be seen between 1931-61 and 1971-81. One of the important features of this growing population was that it was not endogenous by nature but mostly exogenous. The population of Darjiling town grew more than 5 times between 1872 and 1901. From 1901 onwards, the growth has been more or less steady. As it is vivid from the table that only 1931-41, 1951-61, 1971-81 and 1981-91 witnessed some substantial growth. Population increased by 28 per cent, 22 per cent, 35 per cent and 26 per cent respectively. Still this growth was slower while compared to the other towns of the district. This relatively slower growth can be seen in Table 2. As the population growth has been observed from the census record that its more or less steady growth, but the percentage growth of population compared with other towns of the district, Darjiling town's is marginally slow. Among the urban centres in the district, Siliguri witnessed the maximum growth rate in the post-Independence India. It emerged as the important business centre and nodal point of the north-eastern India. Its proximity to Nepal, why all capitals Bhutan and Bangladesh also helped its growth. Kurseong and Kalimpong also witnessed higher growth rate than Darjiling. Immigration was the major factor for growth, as the area flourished and created employment opportunities with the growing tea industry. Nepal provided around 45 per cent of foreign immigrants to India.

Although population in 1991 was 71,469 with 36,912 males and 34,557 females, but overall population seems to be fluctuating due to seasonal immigration and emigration customary to the region. During the tourist seasons from April to June and September to November, thousands visit hill-station and they have influenced these temporary population variations.

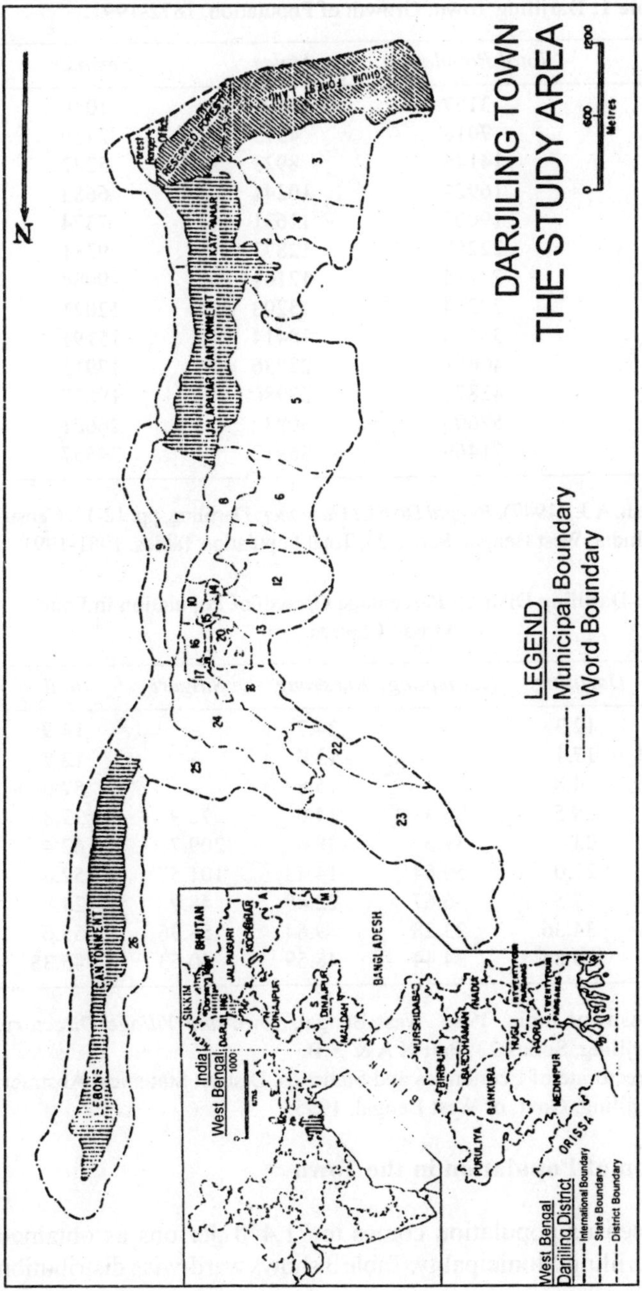


**Fig. 1:** Darjiling Town: the Study Area

**Table 1:** Darjiling Town: Growth of Population, 1872-1991.

| *Year* | *Total Population* | *Males* | *Females* |
|---|---|---|---|
| 1872 | 3157 | 2108 | 1049 |
| 1881 | 7018 | 4579 | 2439 |
| 1891 | 14145 | 8923 | 5222 |
| 1901 | 16924 | 10241 | 6683 |
| 1911 | 19005 | 11631 | 7374 |
| 1921 | 22258 | 12877 | 9381 |
| 1931 | 21185 | 12101 | 9084 |
| 1941 | 27224 | 15203 | 12021 |
| 1951 | 33605 | 18414 | 15191 |
| 1961 | 40651 | 22736 | 17915 |
| 1971 | 42873 | 22986 | 19887 |
| 1981 | 57603 | 30932 | 26671 |
| 1991 | 71469 | 36912 | 34557 |

*Source*: Dash, A.J., (1947), *Bengal District Gazetteer*: Darjiling, pp. 12-13. Census of India, West Bengal, Series 23, Total Population Tables, 1951-1991.

**Table 2:** Darjiling District: Percentage Growth of Population in Four Urban Centres.

| *Year* | *Darjiling* | *Kalimpong* | *Kurseong* | *Siliguri* | *Total* |
|---|---|---|---|---|---|
| 1911 | 12.3 | – | 24.7 | – | 14.9 |
| 1921 | 17.1 | – | 15.6 | – | 12.7 |
| 1931 | -4.8 | – | 15.6 | – | 57.0 |
| 1941 | 28.5 | 36.3 | 14.0 | 72.9 | 33.8 |
| 1951 | 23.4 | 39.5 | 38.0 | 209.7 | 62.4 |
| 1961 | 21.0 | 50.54 | 14.43 | 101.57 | 52.8 |
| 1971 | 5.5 | -6.67 | 22.48 | 48.9 | 24.8 |
| 1981 | 34.36 | 23.28 | 9.64 | 58.36 | 55.6 |
| 1991 | 26.84 | 34.44 | 48.59 | 40.53 | 45.38 |

*Source*: Census of India, 1981. West Bengal, *Town and Village Directory*, Darjiling, Series 23, Part X A & X B.
Directorate of Economics and Statistics. District Statistical Abstract: Darjiling, Govt. of West Bengal, 1995.

## Distribution of Population in the Town

The 1991 census population comes to 71,470 persons as obtained from the Darjiling Municipality. Table 3 shows ward-wise distribution of population between 1971-91 along with ward-wise area to depict more clarity in the study. The ward-wise population growth is positive in most of the wards but some of the wards like X, XIV, XV, XVI, XVII, XIX and XX are experiencing decrease in population growth due to lack of space and location. These wards are in the heart of town and are centres of business and transportation with large number of hotels, restaurants and shops, so there is not any room for

**Table 3:** Darjiling Town: Ward-wise Distribution of Population (1971-91) and Area.

| *Ward No.* | *1971* | *1981* | *1991* | *Area (Acre)* |
|---|---|---|---|---|
| 1. | 1513 | 1761 | 1677 | 90.72 |
| 2. | 1065 | 1404 | 2601 | 162.72 |
| 3. | 2083 | 2373 | 3306 | 374.40 |
| 4. | 1331 | 1668 | 2193 | 99.36 |
| 5. | 2673 | 3536 | 5139 | 275.40 |
| 6. | 2825 | 4339 | 5081 | 101.52 |
| 7. | 1175 | 2164 | 3483 | 148.32 |
| 8. | 1893 | 2705 | 3713 | 72.72 |
| 9. | 2892 | 4203 | 5611 | 92.88 |
| 10. | 1139 | 1322 | 1177 | 13.39 |
| 11. | 651 | 1187 | 2295 | 27.36 |
| 12. | 748 | 1813 | 3456 | 92.16 |
| 13. | 1625 | 2353 | 2547 | 56.16 |
| 14. | 1462 | 1908 | 2471 | 5.76 |
| 15. | 1460 | 1312 | 1008 | 4.32 |
| 16. | 1129 | 1855 | 1353 | 13.25 |
| 17. | 1643 | 1837 | 1793 | 13.25 |
| 18. | 1561 | 1409 | 2440 | 56.02 |
| 19. | 1109 | 2022 | 1082 | 4.32 |
| 20. | 1055 | 1955 | 2264 | 22.03 |
| 21. | 1901 | 1955 | 2264 | 22.03 |
| 22. | 2154 | 3307 | 3052 | 60.48 |
| 23. | 1439 | 2346 | 2408 | 246.67 |
| 24. | 2301 | 2225 | 2851 | 110.16 |
| 25. | 2351 | 2838 | 3733 | 181.87 |
| 26. | 1695 | 2717 | 3109 | 279.65 |
| **Total** | **42873** | **57603** | **71469** | **2615.04** |

*Source*: Census of India, West Bengal, Series 23, 1971, 1981 and 1991 and *District Census Handbook*, Darjiling, 1971 and 1981; Darjiling Municipality for 1991.

expansion. Wards like V, IX, XI, XII, XIII, XVIII, XXIV, XXV and XXVI have shown increase of population between 1981-91. Because these wards fall under the uncovered area of the town where people started settling during this decade. Rest of the wards are having steady growth which are comparatively low. In 1961, there were only 12 wards in the town with an area of 4.08 sq.miles but since 1971, area has been increased to 10.57 km$^2$ with 26 wards.

As regards to the tourist population of Darjiling, it is learnt from West Bengal Tourism Development Corporation that at present about 1,50,000 tourists come to Darjiling every year, of which about 10,000 persons are foreigners. Considering the duration of tourist season of 5 months in a year, the tourist population on an average is 30,000 persons per month. According to the estimation, the maximum number of tourists that can be accommodated per day in Darjiling is 7,460 persons in hotels, holiday homes, rest houses, bungalows, cottages and private houses. The maximum tourist population in a day during peak season has, therefore, been taken as 7,500 persons, which is expected to rise in future. The present population is expected around 84 thousands including tourist population.

**Means of Communication**

The emergence of Darjiling as an urban centre took a long time. It picked up momentum only after the networking of the transportation system. Darjiling as a health resort as well as the progress of the tea industry could not take place without the development of transportation system.

The construction of roads began with the inception of Darjiling. By 1840, a road spiralling and snaking up the hills was completed between *Pankhabari* on the foothills and Darjiling. Between 1849 and 1852, the old military road linking Pankhabari with Kurseong, Dow Hill, Ghum and Darjiling were constructed. The construction of the Darjiling Hill Cart Road was completed in 1865, when, another road between the Tista and Siliguri was being constructed simultaneously.

These roads facilitated direct access for the upcoming settlements, hence, the rate of development along with population started. Vital breakthrough came when the railway line was extended till Siliguri in 1878 and the Darjiling Himalayan Railway came up in 1881, giving conducive atmosphere for growth. The reason behind the improvement in the means of communication was British economic interest and, apparently, it was tea industry which forced the way.

The facilities of roads, rail and electricity had a tremendous effect on the economic and cultural upliftment of the region as a whole. Tracing the genesis of settlement in the region gives a clue to their close link with urbanisation, which has played a decisive role in shaping the life and activities. The emergence of Darjiling town took place in answer to particular needs of the colonial administration, but this could not be a repetitive process conductive to large-scale urban development. This can be ratified from the comparison between Darjiling with other towns so far their growth rates are concerned (Table 4). Even the degree of urbanisation shows that Darjiling and other hill towns are declining in its growth rate than Siliguri because the growth of urban centres is generally restricted due to physical terrain, steep slopes and limited pleasant and sunny days. As such, the economic development is checked by natural landscapes. Thus, the stagnancy occurred in terms of economic development 1951, especially in tea industry and forest management after the Britishers left.

**Table 4:** Darjiling District: Degree of Urbanisation, 1951-1991.

| *Nos.* | *Towns* | *Percentage of Population of Towns to Total Urban Population* | | | | |
|---|---|---|---|---|---|---|
| | | *1951* | *1961* | *1971* | *1981* | *1991* |
| 1. | Siliguri | 6.55 | 10.48 | 12.46 | 15.07 | 16.24 |
| 2. | Darjiling | 6.78 | 6.51 | 5.48 | 5.62 | 5.47 |
| 3. | Kalimpong | 3.36 | 4.02 | 2.99 | 2.83 | 2.91 |
| 4. | Kurseong | 2.36 | 2.15 | 2.10 | 1.76 | 2.00 |
| 5. | Cart Road | – | – | – | 1.08 | 1.02 |
| 6. | Uttar Bagdogra | – | – | – | 0.85 | 0.90 |
| 7. | Jaldhaka Hydel Power Project Town | – | – | – | 0.34 | 0.23 |
| 8. | Bairatioal | – | – | – | – | 0.35 |
| 9. | Mirik | – | – | – | – | 0.53 |
| | Total percentage | 19.05 | 22.80 | 23.04 | 27.55 | 29.65 |

*Source*: Census of India, West Bengal, Urban and Rural Population, Darjiling District, 1981 and 1991.

## Occupational Structure

Darjiling town has grown as a health resort and a centre of trade and communication. The tea industry flourished around the town which has direct impact upon it. But since independence, stagnancy prevailed in tea industry due to *change in ownership* who were trading class not the real tea managers like the Britishers. The new owners had sacrificed quality for the sake of quantity and their main intention was profit.

This stagnancy has led to important changes in the occupational pattern of Darjiling. They are summarised as follows:

1. Proportion of people dependent on plantation has fallen in 1971 in comparison to 1961. In almost all the areas of Darjiling, tea gardens have ceased to become the most important source of livelihood.
2. Small-scale industries have not developed in this area as only 2 per cent are engaged in household industries.
3. No large-scale industries are present.
4. Service has ceased to be a dynamic sector in the economy.
5. In all the towns, rate of growth has fallen.

## Land Use

Darjiling town is complicated and quite restricted to the growth of settlements and development because of landscapes (hills, ridges, spurs and deep valleys) and steep slopes. But despite difficult terrain, it has been witnessing haphazard growth with high rise concrete structures. They have not only spoiled the aesthetic and picturesque view but posed risk with increased probability of landslides. The unplanned rapid growth of town has further strained the already crippled infrastructural facilities and at present the system is on the verge of collaspe.

Land in Darjiling town is divided into two parts on the basis of utilization: covered and uncovered. The covered areas were only for settlement and market, while the uncovered areas were reserved for forests, parks and gardens. Thus, altogether there were 45 per cent-50 per cent of uncovered areas till 1961. But as the number of inhabitants increased slowly with the stagnation of tea industry (migration from tea gardens to town) and immigration from Nepal,

Tibet and other parts of India, the uncovered areas have been slowly cleared. At present, only about 20 per cent of uncovered land is safe, and forested areas are seen in patched at Senchal Catchment Area, Jalpahar cantonment, Birch Hill and Lebong. However, these areas too sparse vegetation cover.

The most picturesque and attractive point of the town is observatory Hill, Chowrasta. The Mall, which commands a magnificent view of the mountains and valleys by which Darjiling is surrounded. The town is projected by ridge which narrows at the top and its eastern slope is very steep: the market (*bazar*) and nearly all the old hotels and houses have been build on the more gentler western slope. There are number of hotels, restaurants and three super markets and numerous shops owned by all ethnic groups. Most of the traders are Marwaris, Biharis, Bhutias and few Nepalese too. The Sadar Police Station, Fire Service Office and two Government Hospitals are located within the market area. The Highway (Hill Cart Road) passes through the heart of market since it's a typical example of linear settlement and is the main characteristics of hilly region urban sprawlling. This is the only highway which is connected with *terai* regions of West Bengal especially Siliguri (gateway of North-east India) but facing lot of problems like waterlogging, traffic jams, etc. due to the increase of shops and hotels which has narrowed the road. There are some parks laid out by the Britishers like Botanical Garden (just below the bus stand), Strubbary park, Governor's house garden, Birch Hill Park, etc. Darjiling is occupied by beautiful buildings, Britishers masterpiece, summer residence of Governor of Bengal whose true copy has been constructed by Maharaja of Cooch Behar at Rose Bank, Eden sanatorium and Lewis Jubilee sanatorium. This town is the place of different ethnic groups so it has most of places of worship: Temples, Monasteris, Churches, Mosques and Gurudwara. The town is fairly well provided with means of recreation. There are a number of open spaces and playing grounds, Gorkha stadium at Lebong, cinema halls, auditorium, Gymkhana clubs for indoor and outdoor games like skating, snookers, table tennis, lawn tennis and different discos and pubs have come up. Many schools and colleges are famous and have well-constructed buildings of architectural merit. The town is well provided with roads and paths, many surfaced with tar macadam and most well fenced but not maintained by both the public and authorities. Growing human number had its impact on the existing ecological systems. In and around Darjiling, there has been too much interference with natural process in the name of making the environment habitable. The mode of making less lofty houses is being given up and concrete buildings are taking their places. In considerate construction of roads without proper care or maintenance, continuous felling of trees without timely replacement are only a few of the visible crimes, man has committed against nature to upset its balance. The destruction of vegetation has reached such a scale that today not a single tree of original stands are found in or around Darjiling. In fact, all the slopes of the hills for kilometres together are scattered bunches which were planted by the Britishers in the mid-ninteenth century. Here, our concern is with the ecological disorder and biosphere problems in the area. As discussed earlier, over zealous and excessive human interference has destroyed the landscape, robbed of the valuable vegetal cover and created ecological disharmony. Forests have vanished in the wake of meeting the rising. The result has been unpleasant with growing intensity of landslides during the rain and paralysing the life style. It is observed that the ecological stress in the mountains has widening destructive impact on other hill stations too.

## From 3Ts to 3Es: A Quest for Sustainable Development

There has been a shift of focus in Darjiling from 3Ts (Tea, Timber and Tourism) to 3Es (Ecology, Environment and Economy). Tea and Tourism earned considerably well, which has gained only to individual entrepreneur and has lost drastically the developmental aspect. These are the chief props of industrial base. Darjiling's pre-eminence as a tourist centre and richly flavoured tea production needs no mention. "Problems of these two industries are many and unless cured may paralyse the economy in the near future."[3]

Timber and tea earlier and tourism of late invited concern from all quarters. Human response to environment in the Darjiling Himalayas attracts large number of tourists, holiday makers, adventurers and mountaineers and many others to the "Queen of the Hill Stations". This has given rise to several insoluble problems. During two peak seasons (March to May and October to November) along with overburdening on the limited resources, water scarcity in Darjiling town has become a major problem. Tea gardens came up at the cost of vegetation on slopes. And unconsolidated felling of trees for local use have further accentuated the problem. This is where the ecological concern has gathered attention. This is where people talk about an economy-based sustainability. Preservation of natural environment and sustainability of the ecological system have gained momentum.

The development problems of the hill areas are qualitatively and quantitatively different from those of plains. Certain basic requirements are to be provided in the hilly areas before we identify areas of priority action. They are:

(a) The Darjiling Hill area can be developed economically only with the planned and scientific exploitation and utilisation of the local resource base. At the same time, efforts should be undertaken to keep a vigil so that these are not over exploited and wasted. An environment-oriented eco-restoration programme is the only alternative.
(b) As a consequence of the natural limitations of agricultural production, hill area development should include alternative sources of employment and income generation other than agriculture, e.g. horticulture, forestry, animal husbandry, tourism, small-scale industry, etc.
(c) In order to stop ecological crisis and avert firewood problem, alternative sources of energy need to be evolved in the hilly areas which can definitely check deforestation.
(d) Considerable improvement in the infrastructural facilities are to be provided.
(e) Furtherr population encroachment on highly sensitive and fragile areas need to be discouraged in order to save environmental deterioration.

In order to keep up the sustainability in developmental activities, corrective actions will have to be taken in some priority areas. Some amounts of environmental degradation are bound to take place if development activities are to be pursued. These corrective steps can arrest further environmental decay and at the same time can encourage development. These steps are essential, in order to save this hill town from further chaos, resulting from land use changes, as shown in Fig. 2.

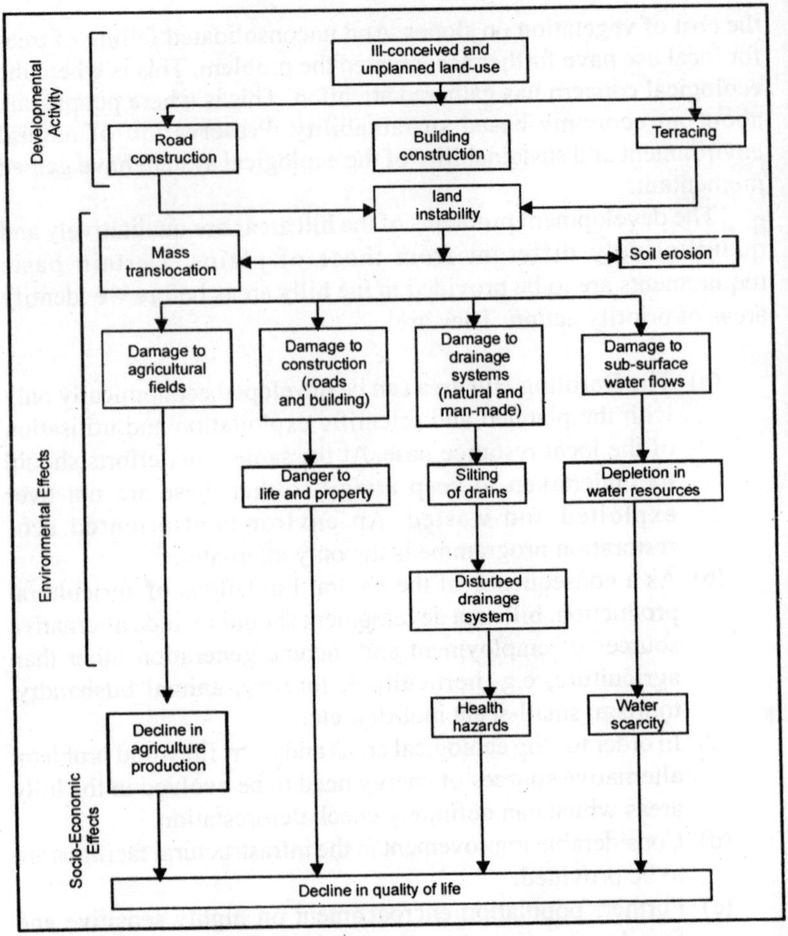


**Fig. 2:** Hill Town: Consequences of Unplanned Land Use.

## NOTES

1. Clark, J. I., (1972): *Population Geography*, 2nd Ed., Oxford Pergamon Press, London, p. 14.
2. O'Malley, L.S.S. (1907): *op. cit.*, p. 35.
3. Chakraborti, P.K. (1989): "The Darjiling Himalaya: A Study of Environmental Resource Management and Development Strategies" in I. Mohan (ed.) *Environmental Issues and Programmes*, Ashish Publishing House, New Delhi, p. 52.

**REFERENCES**

Census of India, West Bengal, Series 23, Total Population Tables, 1951-1991.

Clark, J. I., (1972): *Population Geography*, 2nd Ed., Oxford Pergamon Press, London.

Darjiling Municipality Office Records. Proceeding Books, 1994-95.

Dash, A.J., (1947): *Bengal District Gazetteer: Darjiling*, pp. 12-13.

Basin, S., (1984): "Effect of Modernization on Darjeeling Himalaya". *Geographical Review of India*, Vol. 46, pp. 28-33.

Bhattacharya, B., (1983): "Urbanisation in the Himalayas: Case on North Bengal", in T. Singh and J. Kaur (eds.), *Studies in Eco-Development: Himalayas Mountain and Men,* Print house, Lucknow, pp. 413-24.

Chakravarti, P.K., (1989): "Darjeeling Himalayas: A Study in Environmental Resources, Management and Development Strategies" in I. Mohan (ed.), *Environmental Issues and Programmes*, Ashish Publishing House, New Delhi, pp. 47-57.

Dasgupta, M., (1983): "Underdevelopment and Ecological Balance: A Case Study of Darjeeling Hill Areas", in T. Singh and J. Kaur (eds.), *Studies in Eco-Development: Himalayas Mountain and Men*, Print House, Lucknow, pp. 129-147.

Dasgupta, M., (1985): "Demographic Pattern of the Hill Areas of Darjeeling District: A Study of Census of 1971", in S.K. Chaube (ed.), *Himalayas: Profiles of Modernisation and Adaptation*, Sterling Publications Pvt. Ltd., New Delhi.

Kar, N.R. and Mukherjee, S.R., (1972): "Urban Land Use, Functional Activities and Social Morphology of Darjeeling". *Proceedings of the 21st IGU Symposium on Urban Land Use*, Calcutta, National Committee for Geography, pp. 31-43.

Mitra, S. and Roy, S. (1985): "Changing Occupational Structure in Darjeeling Hills". *Geographical Review of India*, Vol. 47, pp. 61-67.

Munsi, S., (1984): "An enquiry into the Nature of Frontier Settlements: A Case Study of Hill Darjeeling". *Geographical Review of India*, Vol.43, pp.122-137.

Singh, S., (1987): "Influence of Physiography on the Location and Land use of Tea Gardens of Darjeeling District". *Indian Geographer*, Vol. 1, pp. 114-21.

Singh, T.V., (1980): "Impact of Tourism on Environment: A Partial Analysis of Uttarakhand Himalaya". *The National Geographical Journal of India*. Vol. 26, pp. 75-80.

Venkateshvaram, R.J., (1988): "Darjeeling Needs Rapid Infrastructure Growth". *Economic Times*, 15th September, p. 4.

# PART IV

# Socio-economic Development

# 14

# A Study of the Historical and Present Day Changes in Land use Profile, and Their Driving Forces in Mauritius

B. LALLJEE AND S. FACKNATH

## Introduction

Land is a basic natural resource that holds the key to the survival of mankind and sustainable development. Land includes soil, water, vegetation and the microclimate, and provides humanity with the essential life support systems and it meets our ever-increasing demands for food, fodder, energy, settlements and industries. The rapid rise in population, industrialisation and urbanisation has brought considerable changes worldwide in land cover and land use. With the progressive pressure of human activities, land is increasingly being subjected to pollution and degradation in various forms. Land cover refers to the physical characteristics of the earth surface, together with the distribution of vegetation, water, desert, ice and other physical features of land, including those of anthropogenic origins, e.g. settlements, industries, mines, etc. Land use is the intended employment and management of land. Land cover and land use changes may be grouped into two main categories, viz., conversion or modification. Conversion is permanent and refers to changes from one land use or land cover to another, for e.g. removing a crop for construction of industries. Modification refers to a change in land attributes while maintaining the land use, e.g. a residential area being converted to an industrial zone, or land under tea being converted to sugarcane.

Analyses and understanding of historical land use and land cover are very important factors in determining the present and predicting future land use patterns.

## Land Resources of Mauritius and Their Characteristics

The Republic of Mauritius consists of one major island and a group of outer islands and islets, with a total area of about 2045 $km^2$. The islands of Mauritius and Rodrigues are purely volcanic in origin. Mauritius has been shaped by two main lava flows: an older and a younger volcanic series, between which there has been a long period of quiescence and intense weathering. The younger volcanic series comprise two periods of eruption – an early lava flow and a late lava flow. The late lavas have given rise to very rocky and immature soils, and cover nearly 70 per cent of the island.

The main physiographic features of the island are the broken, highly weathered, mountain ranges with associated peaks enclosing a central plateau, and a coastal plain outside these mountain ranges. The parent material of soils in the South-west plain is of colluvial and alluvial origin, which have given rise to fertile soils, such as the Dark Magnesium clays. The natural drainage system of the island is radial, originating from the central Plateau or the central tableland. The rivers are deeply incised, giving rise to picturesque waterfalls and cascades.

### *Climate*

The climate of Mauritius consists of two main seasons, a winter from May to October and a summer from November to April. July is the coolest month, and February the warmest. The island is frequently visited by cyclones or cyclonic depressions from December to February, although the official cyclonic period is from the beginning of November to the end of May. Annual rainfall ranges from 900 mm in the West to 5000 mm on the Central Plateau. There is a succession of climates because of the variations in altitude, and the exposure and distance of the region from the sea. A total of 24 microclimates have been identified which can be grouped into 3 broad classes: the Superhumid Zone, which covers about 40 per cent of the total area, the Humid Zone which covers 35 per cent, and the Subhumid Zone which covers the rest.

### *Soil*

The agriculturally important soils of Mauritius fall into two main groups:

(a) The typical mature ferrallitic soils or latosols, in which the decomposition of the parent basaltic lava rock has proceeded to such an extent that, except for large rounded boulders and stones, there are now no undecomposed minerals in the soil complex.

(b) The typical immature soils, the properties of which are affected by the presence of minerals still in the process of weathering, and which in the field, are characterised by the presence of a more or less high proportion of stones and gravel of vessicular lava. In the USDA system these soils are classified as Oxisols.

The 6 main different soil types important for agriculture are Low Humic Latosols, Humic Latosols, Humic Ferruginous Latosols, Latosolic Reddish Prairie, Latosolic Brown Forest and Dark Magnesium Clays.

### *Water*

Mauritius was, at one time, considered to be a "well watered island" with an annual average rainfall of 2000 mm. Even the subhumid coastal regions, receiving an annual rainfall seldom less than 1000 mm, could be regarded as having adequate water when compared with many countries. However, in the subhumid region the uneven distribution and frequency of rainfall, coupled with high rates of evaporation, gives rise to moisture deficits, and sustained crop production is not possible without irrigation.

In recent years, global climatic changes and other factors have led to a reduction of about 100 mm of rainfall every year (Sok Appadu, pers. comm.); 1999 was a severe drought year, while 2000 brought less than the normal precipitation.

The water resources of Mauritius consist of 4 main underground aquifers, forty seven watersheds and associated river basins. Surface storage of water is mostly confined to natural basins in the central tableland. The Mare-aux-Vacoas is the largest reservoir. The Midlands reservoir, started in 1928, but then abandoned, and restarted in 2000, and completed in 2003, will cater mainly for the irrigation needs of the Northern plains. This is the second biggest reservoir in the island.

**History of Land Use in Mauritius**

Land use on the main island of Mauritius has always been inextricably linked with the economic history of the country. Land use has had a direct and important influence on the economic situation of the island, while the economic conditions in turn have had a direct and important influence on the land use situation in the country.

The uninhabited island of Mauritius was once covered with valuable ebony forests, with most species of trees being endemic to Mauritius. Birds and bats (the only mammals before the arrival of man) carried by errant winds, and a few hardy land animals such as tortoises and lizards, having survived in the sea on rafts and tree branches, established themselves on the island and evolved into unique species. Some tortoise species weighed over 2000 pounds. With very few predators (only the owl and kestrel), the newly arrived birds too evolved into highly unusual and unique forms, e.g. the large flightless Dodo, the Echo Parakeet, the Mauritian Fody, the Olive white-eye.

Once a stop over point in the Indian Ocean for Arab and other trader ships to stock up on fresh water, fruit, meat, fish, and timber, it was colonised by the Dutch in the sixteenth century. The arrival of man in the sixteenth and seventeenth centuries, his need for food, shelter and timber, resulted in dramatic changes to the land cover. The most dramatic change in this point of the island's history was undoubtedly the disappearance of the Dodo bird, which was hunted to extinction by the Dutch.

Major land use change resulted from the introduction, by the Dutch, of sugarcane in the eighteenth century. Grown principally for the production of 'arrack', an alcoholic drink obtained from the distillation of molasses, sugarcane soon assumed a significant role in the economy. More and more of the virgin forest was cleared to make way for sugarcane plantations and associated activities (roads, construction of mills, houses for owners, slaves, stables for horses and cattle, etc.). The French administration that followed the Dutch, gave large acreages of freehold to those of their people who had settled on the island, giving rise to large estate-holders. Imported black slaves from the African mainland provided the labour needed to work in the sugarcane fields. The passage of the island from French to British hands ensured a ready market for Mauritian sugar in British markets, and fixed prices by the colonial powers resulted in a steady expansion of the sugar sector in Mauritius. At this time, more than 100,000 indentured labourers were brought to Mauritius, through the East India Company, from various states of India to work in the cane fields. The Trade Bill of 1825 provided a further boost to sugarcane planters, and sugar production trebled by 1835. New estates were established, new lands were opened up to sugarcane, existing cultivations were enlarged, and productivity was increased in a number of ways. Processing of sugar was mechanised by the introduction of water and steam mills, vacuum pans, centrifugal dryers, all of which helped to make the milling of cane into sugar more efficient and more rapid. However, in order for this new method to be economical, large amounts of cane needed to be transported quickly to the mills and was hence profitable only on estates having greater than 300 hectares and more than 100 slaves and animals.

By 1822, sugar dominated the export market and by 1829, income from sugar exports rose from about 30 per cent to 90 per cent. Mauritius had already embarked on a monocrop economy, with sugarcane cultivation exceeding all the other cash and food crops, such as coffee, cloves, indigo, ebony, maize, manioc, rice. Acreages under sugarcane increased more than 5-fold from 1816 to 1835 (Table 1).

**Table 1:** Acreage under Different Crops (arpents*).

| | *1816* | *1826* | *1829* | *1831* | *1835* |
|---|---|---|---|---|---|
| Wood | 108,418 | – | 91,817 | – | 100,405 |
| Manioc | 17,219 | 15,212 | 12,305 | 10,191 | 6,173 |
| Sugar | 10,221 | 24,967 | 48,485 | 53,892 | 57,933 |
| Wheat | 4,335 | 12,948 | 6,884 | – | 1,807 |
| Indigo | 2,474 | 129 | – | – | – |
| Coffee | 2,161 | 1,060 | 702 | 393 | 434 |
| Cloves | 744 | 949 | 569 | 473 | 62 |
| Ebony | – | – | – | 83,757 | – |
| Others | 8,225 | 12,002 | 6,677 | 5,148 | 4,180 |

* arpent is a French unit, roughly equivalent to 0.422 ha.

*Source*: Compiled from various sources.

***Land Use***

While overall, sugarcane production figures for the whole country increased dramatically, the increase was not uniform throughout the island. Proximity to Port-Louis and easy access and communication methods were determinant factors for the sugar industry. More importantly, suitable soil types, favourable environmental and climatic conditions, and topography were important factors that influenced the type of land use.

Of the 9 main districts of the country, the fertile, suitable conditions of Pamplemousses, Flacq, Riviere du Rempart, Plaine Wilhems and to some extent, Grand Port, had the majority of sugarcane lands. In these regions, sugarcane cultivation expanded rapidly to the detriment of other crops, namely coffee, cloves, rice, wheat, maize, tobacco and manioc. In Pamplemousses alone, acreage under sugarcane increased from 4,586 arpents in 1827 to 17,000 arpents in 1836, while in Plaine Wilhems the figure rose from 4,420 arpents to 7,764 during the same period.

Production of manioc and maize (considered as slave staples), and even fresh vegetables, fell so drastically on some large estates that they came to depend on the smaller neighbouring estates or on the free blacks for these produce. Other estates grew just sufficient manioc and maize to feed the slaves, and turned over the other arable areas to sugarcane.

Moka and Riviere Noire were 'minor' sugar districts, with a few large estates cultivating sugarcane, in combination with manioc, maize and sweet potatoes. Moka concentrated essentially on market gardening and producing fresh vegetables.

Riviere Noire, having a mountainous topography and forested areas, was not suitable for sugarcane. It had cotton cultivations, which dropped from 3,911 arpent in 1817 to only 3 arpent by 1830. Due to its hot and dry climate, Riviere Noire had a large cattle population, mostly imported from Madagascar, and cattle breeding was an important activity in the district.

Savanne, being a heavily forested district, had woodcutting as an important industry.

The coastal regions (e.g. Mahebourg) saw significant amounts of fishing and other related activities, such as collection of sea urchins and sea cucumbers.

Although the sugar boom of this period heralded technological advancements in the milling and manufacturing of sugar, particularly after 1845, this was not reflected in equal measure in the production of sugarcane. While new imported varieties of sugarcane were planted and tested for yield, seed planting introduced, animal-driven carts replaced by a railway system in 1864, overall the fieldwork was still carried out manually by slaves. Estate-owners resisted the use of labour-saving devices, such as the plough. This was partly due to the rough topography of the land, which made the use of farm implements difficult, and partly to ensure that the slaves and the newly introduced indentured Indian labourers were kept busy.

An important consequence of the sugar expansion period was the reorganisation of the slave labour force.

During this period of Mauritian history, it was the availability of labour that was the determinant factor that decided the type of land use.

The sugar industry opened up opportunities for side industries to flourish, and led to the establishment of a new category of traders and small businessmen. These were mostly of Indian and Chinese origin. The numbers of free blacks increased from 42 per cent in 1807 to 68.9 per cent by 1830, and they owned a significant proportion of cattle, pigs, sheep, goats and mules.

Another aspect of the island's land use history relates to what has been termed as the 'agro-commercial bourgeoisie', which grouped the large, Franco-Mauritian, estate-owners, and their entry into the political scene of the day. Essentially of French descent, all these planters were slave-owners and strongly resisted any amelioration or change in the working and social set-up. The sugar planters were the most vociferous and aggressive opponents of slave emancipation. Their interest in maintaining slavery and the existing conditions of slaves was initially rooted in the agricultural, economic and social (slave ownership was evidence of the owner's economic and social status) situation of the times. Their common fight against British efforts to improve the lot of the slaves brought in the judicial and legal element. By 1930, these interests merged and gradually led to the emergence of political groups. Thus sugar planters entered politics. The inevitable result of this coupling of sugar interests and political power meant that the main sugar producing districts produced the most vocal, active and fanatical planter-politicians, and who were the last to accept any kind of change.

### *Land Use and Slavery*

Slavery was established in the eighteenth century with African slaves being brought in mainly from Senegal, Mozambique and Madagascar, to work in the sugarcane fields. The sugar expansion of the early eighties needed large numbers of slaves to clear forests, cultivate, harvest and transport cane, clean and feed the estate animals, as well as for other activities (constructions of roads, houses, as domestics, etc.). Technological improvements meant that slaves were needed to run the new steam and water mills, as boilers, mechanics, burners, potters, coopers, etc. This increased demand for semi-skilled labour led to a reorganisation of the slave force, by importing slaves from other colonies, transferring more slaves to sugar-related work, and most importantly, using unhitherto used labour, i.e. women, children and elderly slaves. Putting slave women to work in sugarcane fields for jobs previously done only by men, was an important step in the socio-economic scenario of the country.

The type and extent of workload of slaves varied from estate to estate and from district to district. The most skilled and able slaves were attached to the sugar factories to run the steam and water mills; less skilled to produce 'arrack' and rum, as market sellers, pig breeders, cattle and poultry keepers, fishermen, and gatherers of sea-urchins, sea cucumbers and molluscs, as well as for copra production and processing.

While the sugar expansion can be considered, economically, to be the brightest period in the island's history, it was the darkest in terms of its social record. Economic factors overrode logic, reason and human sentiment. Fear of slave liberation and the resulting loss of available labour caused planters to adopt extremely repressive laws. The strong and aggressive resistance of the large estate-owners to slave liberation (or even the least improvement in the working and living conditions of the slaves), and the inhuman treatment inflicted on the slaves, made Mauritius the most notorious of the British colonies.

Even among the white population, class distinctions were a result of agro-economic conditions. The large sugar planters formed the elite of society. The less affluent whites were considered a separate class, distinct and excluded from the elite, while the traders, small merchants, soldiers, businessmen formed smaller, separate groups.

***In the Wake of the Sugar Boom***

The sugar boom did not last very long. Since the sugar expansion had occurred against a backdrop of a declining labour supply, inadequate infrastructure, no improvements in production and insufficient ones in sugar manufacture, and an imbalance in the type of agriculture practised (namely, an increase in sugarcane at the expense of food crops), and was followed, in the late eighties, by falling world sugar prices, it inevitably led to a series of serious problems. Lack of sufficient capital to invest in new land, in machinery and slaves, and enforcement of anti-slavery laws, brought many estates to ruin. Being a highly labour-intensive activity, only the large estates, which had large numbers of slaves, could continue to produce sugar. The smaller estates, particularly those under Indian or Coloured ownership, either grew crops other than sugarcane or were engaged in other activities.

One response of the sugar planters to this economic crisis was to consolidate their efforts by merging into companies. The shift from family-owned estates into companies allowed a degree of centralisation, more mechanisation, more efficient and, hence, cheaper production of sugar, influx of capital through bank loans and credit facilities, cheaper housing and other provisions for the labourers, and avoidance of duplication of certain units (e.g. weighing points). The approximately 260 mills that existed in the mid-eighties were reduced to about 80 by the turn of the century.

Another response of the planter to the drastic economic situation was to sell marginal land to their labourers or give them to 'metayers' to farm the less productive portions of the estate land, on the condition that all the cane produced was given over to the estate. The actual process by which the land was parcelled out to metayers and labourers have never been fully studied nor documented.

The selling of unproductive portions of land (called 'morcellement') and the '*metayage*' system was another crucial step in the socio-economic scenario. On one hand, it widened the gap between the large planters and what came to be known as the small planters. On the other hand, it created several more strata in the social ladder, ranging from the millers, estate-owners, large planters, small planters, metayers, labourers, etc. The scission between millers and planters created an additional class category—the middlemen or brokers – who intermediated between the producers and the processors.

Cyclones, pest and disease outbreaks in sugarcane fields, animal disease (epizootic) epidemics (e.g. the Surra disease of cattle in 1902) were some of the factors that affected the process of morcellement from time to time, causing sharp falls in land being parcelled out, followed by upward trends in land sales when conditions became favourable.

The morcellement period also led to the formation of most of the villages and hamlets that exist today.

The difficulties faced by the smaller planters were numerous. The small sizes of their holdings made sugarcane cultivation uneconomical. In addition natural disasters, unfavourable weather conditions, crop failure, high interest rates on loans from moneylenders meant that many lost their land and livelihood. However, an interesting insight into the psychology of the planter of the time is obtained by the fact that in spite of these difficulties, and in spite of the fact that growing vegetables and other food crops was easier and more profitable than growing sugarcane, the small planters continued to convert their vegetable cultivations into sugarcane, perhaps in an attempt to emulate the affluent large planters.

The small planters, the majority of whom were ex-indentured labourers, brought from India, also viewed land as a status symbol, and the way to social success. Land owned was seen as a family asset and land-ownership was considered to be a family affair, and immense sacrifices were made for the education and betterment of the family and of future generations.

Legislation, resale and inheritance among the new land-owners subdivided the land into smaller and smaller, uneconomic holdings, so much so that by the mid-1990s, a small planter was regarded as one cultivating between 0.1 and 4.2 ha. land.

The small planters, which numbered about 35,000 in the early 1990s, formed a group very different from the large planters. They cultivated sugarcane on about 65,500 plots making a total area of 23,634 hectares (Deville, 1990; Anon, 1990), representing about 28 per cent of the total harvested area, but accounted for only 18 per cent of the total sugar. Their productivity was, and is, hence, much less than the estate planters (Lalljee, 1991). Table 2 gives the cane lands indexed according to area group as at 1990.

**Table 2:** Cane lands Indexed According to Area Groups.

| | *Area Groups (ha)* | | | | | |
|---|---|---|---|---|---|---|
| | *0-0.42* | *0.42-0.84* | *0.84-2.1* | *2.1-4.2* | *4.2-10.5* | *Total* |
| Area indexed | 8292 | 5990 | 6120 | 2543 | 692 | 23634 |
| % total area | 35 | 25 | 26 | 11 | 3 | |

*Source*: Deville (1990).

About 90 per cent of the cultivated land in the country at present is held in the form of freeholds and title deeds, the rest is cultivated by the metayers. About 86 per cent of the total area indexed fall within the group < 2.1 ha. and 11 per cent within the group 2.1-4.2 ha. According to a survey (Deville, 1990) the main constraints to increasing productivity of small planters are land preparation and derocking, weediness and gaps in the field, labour, transport, irrigation and socio-economic factors such as theft, distant fields and lack of time.

### *Agricultural Land Use in Recent Times*

Agriculture, today occupies 40 per cent of the total land area (90,000 hectares) of the island, and contributes towards 6 per cent of the GDP of Mauritius. Sugar production has played a pivotal role in agricultural development in the country.

### *Crops*

Sugarcane has, and continues to, dominate Mauritian agriculture (Table 3). The reasons for this situation are many, among them the adaptability of sugarcane to the diversity of local soils and climate, its relative tolerance to cyclones and drought, and technological advancements that has allowed the benefits of monocrop agriculture without many of the associated drawbacks. Land under sugarcane is cropped continuously for 8 to 10 years, after which it is uprooted and the land prepared, mostly by machines, and derocked to help mechanisation. In the virgin plantation, vegetable and other crops such as beans, maize, groundnuts are grown as mixed or interline crops.

**Table 3:** Total Land Area Classified by Utilisation.

| *Land Use* | *Area (hectares)* | *(Area) %* |
|---|---|---|
| **Agriculture** | **90,000** | **48** |
| Sugarcane | 83,625 | 45 |
| Tea | 3775 | 2 |
| Tobacco, vegetables and other crops | 2600 | 1 |
| **Forests, scrub area, grasslands and grazing lands** | **65400** | **35** |
| Planted forests | 6775 | 4 |
| Natural forests | 2387 | 1 |
| Savannahs, grasslands, meadows | 7438 | 4 |
| Scrub and other forest lands | 48800 | 26 |
| **Reservoirs and ponds** | **1180** | **1** |
| **Swamp and rocks** | **1430** | **1** |
| **Roads** | **3465** | **2** |
| **Built up areas** | **25000** | **13** |
| **Total Island** | **186,475** | **100** |

*Source*: Digest of Agricultural Statistics, 1996.

Mauritius produces an average of about 600,000 tonnes of sugar, which is exported mostly to the European Union and the U.S. The sugar industry is run on modern lines and is profitable.

There is close South-South collaboration between Mauritius and other countries in the region, e.g. Mozambique and Botswana in the SADC (South African Development Community), particularly in the agricultural sector. The sugar industry in Mozambique is benefiting from the experience and technical know-how of the Mauritian sugar industry. The year 2001 saw the first sugarcane harvest in Marromeu, Mozambique, which was the fruit of this collaboration.

The tea industry has not been very profitable in Mauritius. This sector has, in fact, been heavily subsidised by the Government in the past. Since this was a particularly labour-intensive sector, it was mainly the severe unemployment problem of the seventies that gave the tea industry its boost at the time. Falling world prices for tea, coupled with the present high cost of labour made this a highly uneconomical industry, and during the last decade, Government encouraged the conversion of tea gardens into sugarcane plantations.

Tobacco is grown in Mauritius mainly for local consumption. In 2001 the total area under tobacco was only 383 hectares (Table 4).

Vegetables and other crops are mainly produced by the small growers within pure stands or as interline or mixed cropping. The interrows of sugarcane (1.7 metres) are used advantageously for the production of short duration crops. Mauritius is generally self-sufficient in fresh vegetables, except in times of cyclones and severe drought.

**Table 4:** Changes of Land Cover under Cane, Tea and Tobacco Cultivation in Mauritius.

*(in hectares)*

| *Crops* | *1990* | *1991* | *1992* | *1993* | *1994* | *1995* | *1999* | *2000* | *2001* |
|---|---|---|---|---|---|---|---|---|---|
| Sugarcane | 82,350 | 82,070 | 81,150 | 79,452 | 77,467 | 76,132 | 72,563 | 73,056 | 73,197 |
| Tea | 2,905 | 2,870 | 3,135 | 3,140 | 3,028 | 2,077 | 671 | 670 | 660 |
| Tobacco | 635 | 625 | 810 | 746 | 696 | 691 | 391 | 391 | 383 |
| Food crops | | | | | | | 6,059 | 7,357 | 7,918 |

*Source*: Digest of Agricultural Statistics, 1996, 2000, 2001.

Table 4 shows the drastic change in land cover over the last decade. Sugarcane acreage has decreased by 11 per cent whereas tea cover decreased by 77.3 per cent and tobacco by 40 per cent. On the other hand, acreage under food crops has been continuously increasing and has reached more than 7,000 ha. by 2001. This figure is still on the rise.

## Driving Forces of Land Use and Land Cover Changes in Mauritius

Some of the driving forces which have contributed, to varying degrees, to changes in land use on a temporal scale are Population, Labour, Climate change, Education, Global, regional and local trends and markets, Technological innovation, Land speculation, Government policies and agrarian reforms, and Migration of people from rural to urban regions.

### *Demographic and Social Changes*

The per capita total land area and per capita arable land area over the period 1851 to 1996 are given in Table 5. The population of Mauritius increased by nearly 99 per cent in the last 150 years. The major increases were between 1851 and 1911 (mainly due to immigration) and between 1931 and 1972 (mainly due to increase in birth rate).

However, after 1972, family planning policies resulted in establishing a balance between birth and death rates. Today, the population growth rate is 1.4 per cent per annum.

It can be observed that there has been a continual decrease in per capita land over time, and with the current population density of 580 inhabitants per sq. km., Mauritius is one of the most densely populated areas of the world. The increasing population needs space for housing, schools, recreation, markets, roads, etc. and this need undoubtedly trespasses on agricultural lands.

**Table 5:** Land Area (ha.) per Person (1851-2003).

| | | *Ratio (ha/person)* | |
|---|---|---|---|
| *Year* | *Population ×1000* | *Arable + Non-arable* | *Arable* |
| 1851 | 180.8 | 1.03 | 0.57 |
| 1871 | 316.0 | 0.59 | 0.32 |
| 1891 | 370.6 | 0.50 | 0.27 |
| 1911 | 368.9 | 0.50 | 0.27 |
| 1931 | 393.2 | 0.47 | 0.26 |
| 1952 | 681.6 | 0.27 | 0.15 |
| 1972 | 826.2 | 0.22 | 0.12 |
| 1992 | 1126.2 | 0.26 | 0.09 |
| 1996 | 1133.7 | 0.17 | 0.08 |
| 2003 (estimated) | 1300 | 0.15 | 0.07 |

*Source*: Lalljee and Osman, 1997.

To satisfy local as well as export markets in terms of agricultural produce, greater and greater proportions of marginal lands (e.g. on slopy mountain sides, wetlands, etc.) have been brought under cultivation. This has led to land degradation, which is a very complex phenomena, apparent only in the long-term.

The movement of people from rural to urban regions (and now increasingly, from urban to rural regions) has also contributed to the changes in land use and land cover changes.

### *Changes in the Labour Market*

Qualitative and quantitative changes in the labour force are major contributors to land use change. To cater for the increased unemployment, just before the country's independence in 1968, a vast programme was developed to set up and encourage labour-intensive industries. The labour force showed a marked preference to move from the open fields into factories and industries. This caused a shortage of manpower on agricultural lands, and the small planters, in particular, bore the heaviest brunt of this phenomenon. In addition, with the introduction of free education in 1976, larger and larger numbers of educated Mauritians aspired for white-collar jobs. Climatic factors, such as cyclones and long periods of drought (e.g. 1997-98), have further discouraged many agricultural entrepreneurs and workers.

The overall result of these situations was that agriculture began to be neglected in the country, causing many to shift their interest to other sectors of the economy.

As can be seen from Table 4, the total acreage under sugarcane has been constantly declining over the years, mainly due to abandonment of land by the small planters. Family inheritances have fragmented their land into portions too small to be economically viable; the land is then left fallow or put under other cultivation, such as vegetables.

### *Technological Innovations*

High yielding varieties of crops, derocking, mechanisation, drip and centre pivot irrigation systems, etc. are some of the technological innovations which led to an increase in the profitability of the agricultural sector, and partly offset the shortage and increasing cost of labour. However, it has been observed that these innovations benefited the big planters and estate planters more than they did the small farmers.

### *Land Values and Land Speculation*

During the past 20 years, Mauritius has witnessed a mushrooming of estate agents. Many of these agents have been responsible for luring people into selling their agricultural lands. Property developers then step in and subdivide the lands into smaller plots for residential and other purposes. It has been observed that many of the present day 'morcellements' have taken place on fertile agricultural lands.

The Government, in 1984, passed the Sugar Industry Efficiency Act, which empowers the Minister of Agriculture to control conversion of cane land to non-agricultural purposes, and which puts restrictions on the subdivision of agricultural land for other purposes. This has helped to curb, to some extent, the trend of land conversion.

### *Other Government Policies*

Government had developed several encouraging and interesting schemes and legislation to revert these lands to sugarcane cultivation, to control land conversion to other land uses, to prevent speculation of agricultural land, and to ensure production at a level that would meet all commitments. The creation of the Farmers' Service Centres (FSCs) in the various agroclimatic zones of the island, and especially the Land Area Management Unit (LAMU) projects provided the necessary help to small planters in order to encourage them to continue cultivation of sugarcane.

Other Government policies include guaranteed prices for selected commodities or price liberalisation (e.g. for potato and onions) and the setting up of regulatory and advisory bodies in order to provide incentives and to encourage planters to diversify production and also to increase yields.

At the end of 1995, the Government negotiated an extra export quota of 85,000 tons of sugar from the EU. In order to achieve this, one of the policy decisions was to uproot tea in order to plant sugarcane. The land in these areas is undulating, and generally situated in the high altitudes, and in regions of high rainfall (3000-5000 mm/year). These soils are naturally poorly drained, and being located on slopes, are susceptible to erosion by water. The cation exchange capacity of these soils being low (Lalljee, 1987) the risk of nutrient leaching is aggravated.

Proponents of this policy of such land conversion were of the view that this would present no problem since all the cultural operations, such as selection of cane varieties, date of plantation, spacing of furrows, fertilisers and pesticide applications would be scrupulously observed, and the ecological aspects strictly monitored. A total of about 2,600 areas of tea gardens was thus converted into sugarcane fields.

Preliminary work carried by the senior author has in fact shown significant reduction in soil fertility after the first year of conversion, with important yield reductions in subsequent years.

***Latest Agrarian Reforms/Global Trends and Markets***

In response to globalisation and changes in the world market, Government has recently released the Sugar Sector Strategic Plan (2001) and the Non-Sugar Sector Strategic Plan (2003). The former envisages a reduction in the number of sugar mills to five, permission to sell/convert part of the sugar land to non-agricultural purposes on condition that the proceeds of the same are reinvested in the sugar sector to further modernise it. This Plan has also introduced a Voluntary Retirement Scheme (VRS) which accords laid-off workers of the sugar industry small portions of land of 7 perches (300 sq. m.) each for residential and/or entrepreneurial purposes. Furthermore, Government's Illovo deal of 2000 made it possible for the Sugar Investment Trust (which includes the small planters as shareholders) to buy 7000 arpents (approx. 2500 ha.) of sugarcane land from the large sugar estates, and to sell it to small planters, sugar industry workers and the general public at reasonable prices, and with credit facilities, with the aim of increasing ownership. The deal also includes conversion of about 200 arpents of sugarcane land to non-agricultural use, without having to pay the associated land conversion taxes. Part of the ex-cane land thus liberated has already been given over for the construction of a cybercity in the region of Ebene.

The Non-Sugar Sector Strategic Plan aims, on one hand, to change the existing perception and practices of present-day agriculture, and to modernise Mauritian agriculture and make it more professional, and on the other hand, to make Mauritius a major agro-processing hub in the Indian Ocean region. In line with this aim, Government is encouraging Organic Agriculture, Hydroponics and Protected Cultivation, high-tech agriculture using latest Biotechnological methods, in order to generate high value added, novel and specialised agro-products for the local, tourist and export markets, and also to help reduce the negative impacts of conventional agricultural practices, e.g. excessive use of pesticides, fertilisers on the environment and on human health.

Similarly in order to give a boost to the tourist industry (which is an area of major economic importance) rezoning of areas has been done. In some of these cases, fertile agricultural land has yielded to tourism, and bungalows and hotels have been constructed in place of cane stalks and vegetation. The latest policy measure, known as the Integrated Resort Scheme, concerns the conversion of more than 3,500 arpents for construction of luxury villas, for sale to expatriates. All of these measures will have important consequences on the land use pattern in the country.

## Major Impacts of Shifts in Land Utilisation

The changes in land use are accompanied by alterations in soil physical, chemical and biological properties. If these alterations are in the direction of sustainability, then the land use change may be a welcome strategy. However, most land use changes are driven by factors that rarely take soil health into consideration, and usually become responsible for land and water degradation.

### *Soil/Land Degradation*

Soil/land can be degraded in various ways, for example, soil erosion, loss of organic matter, increasing soil acidity, contamination by toxic chemicals, salinisation, etc.

Soil erosion is one of the most important soil/land degradation problems in Mauritius.

Mauritius has about 18,400 hectares of land (= 1 per cent of the total area) that are susceptible to soil erosion, caused mainly due to agricultural activities, deforestation and construction on mountain slopes.

Organic matter in soil is, in effect, the site of all life and all the biological activities in the soil. Any loss in top soil results in consequential loss of soil organic matter. The problem in Mauritius is aggravated by certain agronomic practices, among them burning of sugarcane before harvest, which leads to depletion of soil organic matter. Wildfires, accidental or deliberate, further aggravate this problem.

The addition of nitrogenous fertilisers, especially ammonium sulphate, to large tracts of sugarcane plantations, for over a century now, has made Mauritian soils increasingly acidic, and this another cause of soil/land degradation in Mauritius. Acidic soils, generally, are much less productive than are normal soils.

It is only recently that farmers have been advised to add neutralising agents such as lime, sand or even cement to their soil.

Salinisation is yet another land degradation problem of the arid regions. Poor quality irrigation water is one cause of increased level of salts in the soil. In some areas of Mauritius, e.g. Baie-du-Tombeau, the excessive extraction of water from bore-holes has led to high salinity of the water, due to sea water intrusion.

It is well-known that greenhouse gases such as nitrous oxide, carbon dioxide, methane are liberated from soil. The soil can act as a source of, or a sink for, environmental pollution. No work has yet been done in Mauritius to quantify such emissions and determine the existence/extent of the problem.

The agricultural situation in Mauritius (large number of pests and diseases, subsidies on agrochemicals, public demand for perfect, blemish-free produce) encourages farmers to make heavy use of fertilisers and pesticides. Precision farming is practically non-existent and farmers usually adopt a blanket recommendation. This contributes to overall degradation and eutrophication of water bodies. The effects of eutrophication are already apparent in some regions of the island, e.g. the La Ferme reservoir. Coral death and coral bleaching are other serious consequences of poisoning by pesticides and other toxic chemicals.

**Deforestation**

Only 1 per cent of the original forests is still present today. The present total area of forest land is estimated at only 56,403 hectares (Anon, 1992), of which only 21,883 hectares are state owned, the rest being privately owned forest lands.

Table 6 lists the present-day forest areas of Mauritius.

Deforestation for the purpose of agriculture, human settlements and exploitation of forest products has resulted in significant erosion of the fertile topsoil (sheet erosion) in the slopy areas. Furthermore, most of the eroded soil ultimately reaches either the lagoon or reservoirs, or more often, the natural river basins. Siltation of river basins has affected the aquatic ecosystem in many parts of the country, and has caused disappearance of fish and shrimp stocks and populations of other aquatic species.

***Loss of Biodiversity***

Plant and animal life in Mauritius was endangered right from the time humans settled on the island. The ebony, Mauritian Kestrel, Echo Parakeet are threatened with extinction, while several species have already been eliminated (Dodo, giant flightless Black Parrot, Dugong,

**Table 6:** Present-day Forest areas of Mauritius.

| *Nature of vegetation* | *Area (ha.)* | *Total area (ha.)* |
|---|---|---|
| **State Forest Lands** | | |
| Plantation | 11,916 | 11,916 |
| *Nature reserves* | | |
| A. Mainland | 4,018 | |
| B. Islet | 567 | 4,585 |
| Unplantable, protective or to be planted | | 4,710 |
| | | 21, 211 |
| *Pas Geometriques* | | |
| Plantations | 211 | |
| Leased for grazing and tree planting | 230 | |
| Unproductive or rocky | 211 | |
| | | 652 |
| **Privately owned forest lands** | | |
| *Reserves protected by law* | | |
| Mountain reserves | 3, 800 | |
| River reserves | 2,740 | |
| | | 6,540 |
| Forest lands including scrub and grazing lands (estimated) | | 28,000 |
| | **Grand Total** | **56,403** |

Snapper Fish, Parrot Fish). The accidental or deliberate introduction of invasive exotic plant species, such as the Chinese guava, privet and scrambling bramble, as well as mammals such as monkeys, deer, pigs and mongoose, has on one hand, led to the dwindling, and in some cases extinction, of the vulnerable lizards, snakes, tortoises and birds, and on the other hand, increased the already heavy pressure on the remaining 1 per cent of the native vegetation. The Mauritian Dodo is today the very symbol of extinction. There is very little of the original biodiversity left. A few of the endemic palms (Hurricane Palm, Bottom Palm, Royal Palm, Latania Palm) and reptiles (the Telfair skink, Guenther's Gecko, the Boa,) species are now to be found on only 1 of the islets around Mauritius.

In Rodrigues, destruction of trees and habitats to make way for agriculture, excessive grazing by free-ranging livestock and altered species-relationships due to introduced crops and animals, has almost completely changed the original biodiversity and threatened the remaining few endemic plant and animal species on this island.

Today, Mauritius ranks high on the priority list of countries with respect to conservation issues. It comes third, after Hawaii and the Canary Islands, as far as the number of threatened plant species is concerned (IUCN, 1984).

Present day conservation strategies include (i) the setting up of Conservation Management Areas, which are either fenced in and/or their contained exotic and invasive plant species selectively uprooted, and (ii) the *ex-situ* propagation of endemic plant species difficult to regenerate in the wild. Faunal conservation includes captive breeding programmes which have succeeded in bringing the Mauritian Kestrel, the Pink Pigeon and the Rodrigues Fruit Bat back from the brink of extinction, while recently the café marron (*Ramosmania rodriguesii*), which had only 1 individual left in Rodrigues, has been successfully propagated at the Royal Botanical Gardens in UK and transplanted in Rodrigues. Species less amenable to captive breeding, such as the Echo Parakeet, are being managed in the wild by protecting their eggs and chicks from rodent predators, and their nests and nesting sites from other birds.

## Pollution

### *Fertiliser and Pesticide Use*

Mauritius is listed as one of the countries with a high per hectare use of fertilisers – about 139 kg. nitrogen per hectare (Lalljee, 1996). Nitrates in ground water are reaching alarming levels – 36.6 mg /L (Gopaul, 1995). A disturbing fact is that only about 80 per cent of the nitrogen fertiliser applied, especially to sugarcane, is recovered by this crop (Ng Kee Kwong, 1995). Studies at the University of Mauritius (Lalljee, unpublished) have indicated a high percentage of leaching of nitrogen from fertilisers, which may eventually find its way into the groundwater or into rivers, where it can cause eutrophication. Studies in 2003 have shown that some boreholes have quire high levels of nitrate and phosphate pollution, with high levels of microbial contamination by pathogenic micro-organisms, presumably due to intense pig breeding activities in the immediate neighbourhood.

Coral reef destruction in a number of places in the lagoon is believed to be due to nutrient discharge from land. However, published documentary scientific evidence is lacking to support these observations.

About an average of 1100 tonnes of pesticides are applied annually to the arable land, which places Mauritius quite high in the list of countries as regards intensity of pesticide use. This also represents a serious potential risk to the environment and to human health. A large proportion of this amount is in the form of herbicides, applied to the cane plantations, while insecticides are used essentially in the vegetable and fruit sector. Surveys and studies carried out at the Faculty of Agriculture, University of Mauritius, have highlighted the incorrect use of pesticides among farmers, and the resulting contamination of soil, water and harvested crop produce by pesticide residues.

Policy changes, research by the various agricultural institutions of the country, dissemination of this information, farmer education through talks, visits and training sessions, etc. demonstrate the nation's commitment to reduce agrochemical use, and in general, to make Mauritian agriculture more sustainable and environment-friendly.

It is imperative that alternatives to the synthetic pesticides and fertilisers be developed. Rock dust, ash, vinasse, scum are some of the sugar industry by/waste products which can be used as fertiliser supplements and/or soil amendments (Lalljee, 2000). Several of the local plants have shown potential as pesticides and can be integrated with other control methods for the development of IPM programmes (Facknath, 1996; Facknath and Lalljee, 1998, 2000).

### *Other Pollutants*

Sugarcane burning to get rid of the trash prior to harvest is a common practice in Mauritius. Nearly half of the harvestable area is burned annually (McIntyre and Hardy, 1989). Cane burning is resorted to for economic reasons, but it does cause pollution of the environment and has negative results on the soil and yield of cane. According to de Oliviera *et al.* (1994), the soot and smoke in the atmosphere persist for several months.

Discharge by some sugar factories of large volumes of effluents, namely, hot condensed water, carbon column washing, fly ash, scrubbing liquid, alkaline cleaning water and other organic solvents with a high biological and chemical demand affect aquatic life. The populations of various aquatic species in streams and rivers have been severely reduced presumably on account of this. It is only recently that the Ministry of Environment has passed legislation defining absolute parameters for the common effluents.

Municipal wastes, containing sewage and solid refuse, are in some places discharged on land without adequate pre-treatment. In many places (e.g. at La Brasserie, Solferino), solid wastes are disposed of in open dumps.

Industrial wastes containing hazardous substances, such as chromium dyes, add to the pollution problem. The main problem facing water resources in Mauritius come from effluents from dye houses, and processing industries. The recent case (November, 2001) of massive fish and fingerling deaths in the bay flowing discharge of wastewater from a textile dyeing plant is a case in point. Other non-point sources of contamination are sewerage waste water from sugar mills, hotels, agricultural runoff, infiltration from solid waste dumps and excessive fertilisers and pesticides used in certain areas (Anon, 1994b).

Biochemical decay of organic materials results in evolution of gases and liquids. The dissolved material moves out due to the concentration gradient. Among the major products of biodegradation are nitrate, methane, carbon dioxide and other chemicals of high BOD (Biological Oxygen Demand) and COD (Chemical Oxygen Demand). The organic and inorganic constituents in the wastes are becoming a major source of land pollution and groundwater contamination. Recently it was argued that the temporary disposal of municipal wastes at Mount St. Pierre in the west of the island, has resulted in serious pollution of the potable water in the region. The planned landfills at Mare d'Australia has not yet been cleared, while the recently operational Mare Chicose is already stated as being environmentally-unfriendly. Experts differ about the suitability and sustainability of these landfills.

While indiscriminate discharge of wastes on land is of serious environmental concern, it needs to be recognised that land is the best available sink for the ultimate disposal of wastes. What is needed is pretreatment of wastes, such as composting or compaction. Minimising the wastes at source is the best solution.

### *Destruction of Wetlands*

Wetlands play an important role in an ecosystem, as a buffering zone, as breeding grounds for birds and other fauna, as habitat for plants and animals, etc. The use of wetlands, particularly those in coastal regions, for other purposes, e.g. construction of hotels near beaches, can have serious repercussions on the wetlands and consequently on the neighbouring ecosystems. Since, the negative effects are slow, and apparent only in the long-term, very often the damage is already done before any corrective action is envisaged. Hence, it is important to conserve and protect the remaining wetlands. It is imperative to highlight this problem to policy-makers, as well as the general public and to make an inventory of all existing and converted wetlands in the country, to be followed by appropriate corrective/remedial action.

## Conclusion

Land use changes in Mauritius have essentially been driven by the sugar industry. The vagaries of the global sugar market have impacted upon the local scenarios of land utilisation. Associated with this main economic factor are the subsidiary, contributing factors such as local and global policy issues as well demographic and social changes.

The ongoing and proposed changes in the agricultural scenario of the country can only be implemented successfully with a new nation-wide vision for research and development, and training and capacity building, in new and sustainable agricultural technologies. Beginning with the collation and/or generation of baseline data, leading to research and development into alternative methods and products of agriculture (and other industries), and completing with training and education of all stakeholders, is the way forward to a clean, healthy and sustainable environment.

Although it is true that the past history of Mauritian biodiversity has not been laudable, the present generation is more enlightened and aware of the importance of preserving our environment. 'Productivity with Sustainability' is the present-day motto.

Agricultural diversification, reforestation, creation of nature reserves, captive breeding of endangered species, cleaning up of the environment in general, landscaping are some of the activities which are receiving the most attention. Coupled with public education at all levels and backed by appropriate legislation, this will help to correct past mistakes and to set up Mauritius as a model in this part of the world.

## REFERENCES


Anon, (1990): *The land index project for small planters*, MSIRI, Mauritius.

Anon, (1992): *State of the Environment in Mauritius*, Ministry of Environment and Quality of Life, Republic of Mauritius.

Anon, (1994): *Annual Report*, Tea Board, Mauritius.

Anon, (1994b): *Capacity Building in Environmental Management*. Ministry of Environment and Quality of Life, and UNDP, March-April, 1994.

Anon, (1996): *Digest of Agricultural Statistics*, Mauritius.

Arlidge, E.Z. and Wong You Cheong, Y. (1975): *Notes on the Land Resources and Agricultural Suitability Map of Mauritius*. Occasional Paper No. 29, MSIRI.

De Oliveira, O., Urquiaga, S., Boddey, R. (1994): 'Burning Cane: the Long Term effects'. *Int. Sugar Journal,* 96 No. 1147 E.

Deville, J. (1990): 'The land index project for small planters'. *PROSI*, No. 262. 6-12.

Facknath, S. (1996): 'Efficacy of Neem and Melia in overcoming tolerance of *Plutella xylostella* to some commonly used synthetic insecticides'. In: Neem and the Environment (Eds. R.P. Singh; M.S. Chari; A.K. Raheja; W. Kraus). *Proc. World Neem Conf.*, Bangalore, India March 1993. Oxford & IBH Publishing Co. Pvt. Ltd., 505-515.

Facknath, S. and Lalljee, B. (1998): Research on Allelopathy in Mauritius. In: *Allelopathy Update*. Vol. I: International Status (Ed. S.S. Narwal). Science Publishers Inc., USA. 271-293.

Facknath, S. and Lalljee, B. (2000): 'Allelopathic strategies for environmentally-friendly crop protection'. In: *Allelopathy in Ecological Agriculture* and *Forestry*. (Eds. S.S. Narwal; R.E. Hoagland; R.H. Dilday; M.J. Reigosa). Kluwer Academic Publishers, Netherlands. 33-46.

Gopaul, A.K. (1995): *Eutrophication of La Ferme reservoir*. Report submitted to the Ministry of Environment. Republic of Mauritius.

IUCN, (1984): *The World Conservation Union for Nature*/WWF, 1984.

Lalljee, B. (1987): *Annual Report of the Agricultural Services,* Ministry of Agriculture, Fisheries and Natural Resources, 1984.

Lalljee, B. (1991): *The sugar industry, the Farmers' Service Centres and the Small Sugarcane Planter*. Thesis-Diploma in Management Studies, University of Mauritius.

Lalljee, B. (1996): 'An overview of agricultural pollution, with special reference to Mauritius'. In: Proc. UNESCO Experts Meeting for SADC Countries, Mauritius, 25-30 January, 1999. 115-123.

Lalljee, B. (2001): 'Rock phosphate as an alternative to soluble phosphatic fertilisers'. *Proc. 4th Annual Meeting of Agricultural Scientists* (AMAS), Mauritius. October 1999. (Eds. J.A. Lalouette; D.Y. Bachraz; N. Sukurdeep). Food and Agricultural Research Council, Mauritius. 47-54.

Lalljeé, B. and Osman, A.(1997): 'Land use in Mauritius: An Overview.' In: *Regional Land Cover Changes, Sustainable Agriculture and their Interactions with Global Change*. (Ed. V. Ravichandran) COSTED, ICSU UNESCO. 36-47.

McIntyre, G. and Hardy, M. (1989): 'The influence of cutting methods and burning on cane and sugar yield.' *Sugarcane* No. 1.

Ng Kee Kwong. R.K.F. (1995): 'L'engrais azote a travers le goutte a goutte.' *PROSI*. No. 312. Mauritius.

# 15

# Resource Capacity, Growth and Development

AMAR S. GULERIA AND BHUPINDER S. MARH

## Introduction

Growth denotes quantitative changes, while development includes both quantitative and qualitative change in economic and non-economic parameters. Many policy-makers, professionals and academicians focus only on the conventional approach to development rather than on the development that takes account of environmental impacts. At present, development must address poverty, unemployment undernourishment, inequality, social unrest. What follows is an attempt to examine the concept of limits to growth, estimate the capacity of resources and suggest environmentally safe and sustainable development?

## Limits to Growth and Resource Limits

Whatever limits the growth in size of an individual or in number of a population is considered as limiting factor to that individual or population. Ecological principle of limiting factors is defined as the process of an organism or a group of organisms which depends on a complex of conditions. Any condition that exceeds the limits of tolerance is said to be a limiting condition or a limiting factor[1]. In 1840, this concept, while studying the growth of a plant which is dependent on the amount of foodstuff in minimum quantity, was also referred to as the law of minimum[2].

Limiting factors can be divided into: (i) physical and (ii) biological. Physical factors that limit population growth include factors of climate and weather, the absence of water or presence of excess water, the availability of essential nutrients, the suitability to the terrain and so on. Biological factors involve competition, predation, parasitism, disease, and other interactions between or within species that are limiting to growth. Limiting factors may further be classified into those whose operations are dependent upon the density of population (the number of individuals per unit of area) and those that heave not relation to density[3]. The density-dependent factors are those that increase in their intensity, that have greater effects, or that affect more individuals as the population density increases. Thus, the availability of food-grass and other herbs may be limiting factor for the increase in numbers of domestic cattle in a pasture. The higher the density of cattle, the less grass there is per cow and greater number of cows suffer from food shortage. By contrast, a flood sweeping through the pasture would be a density-independent limiting factor. It would wipe out all the cows whether there were two or a hundred in the area. Density-dependent factors are important in population studies because of their more general and constant operation. They are usually the factors that set absolute limits to growth, that determine the number of individuals that can be supported by the carrying capacity of the area. They are factors which operate to decrease the individual's well-being in a population, that approaches the limits of carrying capacity of its environment. In crowded human population in many parts of the world we see density-dependent factors in operation.

Carrying capacity refers to the pressures of human and livestock population that limit the growth. It refers to support or providing the means of subsistence for a population. It provides the number of persons and animals to be supported per unit of land safely—without causing degradation of resources/environment. Carrying capacity is not fixed. It can be and has been increased steadily over many centuries—through improved means of food production, more efficient means of food collecting (such as commercial fishing), elimination of diseases, improved shelter and sanitation, and the greater availability of energy resources. It is, thus, believed that carrying capacity can continue to be increased which forms the basis for belief that populations can go on increasing. This was true in Europe where industrial revolution increased the availability of food and other materials as a result of which carrying capacity greatly increased. Similarly, with the process of demographic transition, from high death and birth rates, through lowered death rates and rapid increase, to low death and birth rates, the limit of carrying capacity has changed. Therefore, economists believe that industrialization is the answer to the population problems of the third world[4]. But potential exploitation of resources and the industrialization would not necessarily mean prevalence of suitable/healthy environmental/ ecological conditions and it will threaten the lives of millions. Hence, carrying capacity should ensure sustainable development[5].

Thus, carrying capacity of an area is defined as the number of persons and animals that can be carried on it and kept in good condition without damage to the environment and ecosystems. Carrying capacity depends on the soil and climate, the type of native vegetation, and the ability of the vegetation to hold up under grazing and cultivation for food and forage and forage supply. The number of animals to be carried safely on the area depends on the surplus of leafage and stems put on by vegetation. If too many animals are held, not only the surplus is, eaten but also the reserve portions of the plant. If this process continues or is repeated for long, plants weaken and die.

It is estimated that acres needed to support one cow or five sheep for one month vary between 0.75 for an area in excellent condition (90-1000 per cent potential forage yield), 1-2 for a good area (75-90 per cent potential forage yield), 2-3 for a fair land (50-75 per cent potential forage yield), 3-5 for a poor land (25-30 per cent potential forage yields), and 5 + acres for a very poor land (having relative potential forage yields of 0-25 per cent)[6]. Therefore, a rancher who keeps his ranges close to excellent condition will be able to carry more livestock and make greater profits than a rancher who allows his ranges/area/land to deteriorate. However, range condition is a measure of the degree to which a particular type of range approaches its maximum potential yield of forage. It is something quite apart from range and soil type. Thus, a well-watered prairie range with tall climax grass on deep soil, rated in excellent condition, would have higher carrying capacity than a range in excellent condition in the desert grassland region.

The number of animals present at any one time depends on the balance between two forces: (i) biotic potential, or maximum rate at which as species can increase if unchecked, and (ii) the environmental resistance. The sum of all the forces that cause death or lower reproductive gains effects this number. If the environmental resistance is lowered, in one way or another, animal population would increase or the number of animals supported per unit of land or carrying capacity would increase.

Every habitat, for wildlife as well as for domestic livestock, has its carrying capacity, which sets limits on population increase. No wild animal population can be maintained permanently at a level above the carrying capacity, which is determined by the available food, cover, water, and other essentials of life. Yet each year, a population at carrying capacity will produce young. This crop of the young will represent, therefore, an excess above what the environment can support. Either the young must perish or older animals must die to make a place for them or the natural reduction causes must bring reduction in their population by creating their enemies to maintain the ecosystem, Imbalance would disturb the ecosystem. However, carrying capacity of animals will fluctuate from time to time, from place to place, from species to species, from month to month, and from season to season.

Since there is no basis to estimate the total carrying capacity of the world for man and animals, it is stated that the world has 5 billion people as the carrying capacity. Further, any population growth on the islands of Barbados, where over 1500 people occupy each square mile (or 6 persons per hectare), is potentially dangerous to all of the inhabitants. By contrast, Zambia or Zaire in tropical Africa could support a relatively high rate of growth, if resources and land were equitably divided among the people.

However, we made a similar attempt to estimate the carrying capacity in Himachal Pradesh (India)[7]. The limits of carrying capacity in the study areas varied from 190 to 271 persons per sq. km. (or 2-3 persons per hectare). At the same time, around 20 per cent cattle population was found to be surplus. It has been noted earlier that acres needed to support one cow or five sheep for one month based on forage yields vary between 0.75 for an area in excellent condition, 1-2 for a good area, 2-3 for a fair land, and 3-5 for a poor land, and 5 and more acres for a very poor land. It is emphasized that the concept of carrying capacity in each resource region in more relevant in ecological and environmental studies to ensure sustainable development. Carrying capacity can be used to determine the boundaries of a feasible sustainable development. Some of the other features of a resource base that are useful in this context are exhaustibility, potential depletion, multi-functionality, stability, private *versus* public ownership, and closed and open resource systems.

**Sustainable Development**

The emergence of global environmental change adds to appeals for sustainable development. The concept of sustainable development ranges from continued growth to steady, mainly to compensate for the environmental depreciation. The sustainability debate has created a great-deal of concern in recent years. Limits to growth is therefore a process of change in which exploitation of resources, directions have been investments, orientation of technological development, an institutional changes are made consistent with future as well as present needs. It is said that sustainability takes for granted a balance between economic development (all quantitative and qualitative changes in the economy that offer positive contribution to welfare) and ecological sustainability (all quantitative and qualitative environmental strategies that serve to improve the quality of an ecosystem which have a positive impact on welfare). Sustainable development will require integrating environment and development policies, meaning that activities formerly guided by purely economic criteria will now have to satisfy conservational criteria as well. It needs to control the negative changes which are a consequence of current patterns of resource use that over-exploit the resources. An important concept in sustainable development is the carrying capacity which indicates the number of persons and animals that a region can support under the constant and dynamic technologies without degradation of resources and environment.

Sustainable development is considered as "development that meets the needs not the greed of the present without compromising the ability of future generations to meet their own needs.[8] In the context of agriculture it entails strategies that augment existing land and water resources through investment, technologies to increase the productivity of these resources, and pricing policies as well as management practices that economize on the use of such inputs.

Land-saving technological change represented by the Green Revolution was the single most important factor contributing to sustainability. Extension of cultivation to marginal lands and to area under forest has been halted. The amount of fallow land highly correlated with area irrigated, implying that the progress in yield-increasing technology has reduced pressure on marginal lands. The rapid displacement of bullocks by tractors and cows by buffaloes as milch animals in prosperous irrigated areas also indicates a similar trend. But the concentration of cattle among land-scarce small holdings and the rapid increase in the number of small animals like sheep and goats are exerting pressure on environment.

## Strategies for Sustainable Agriculture

To bring agriculture on the sustainable development path, effective strategies are required. In the first place, there is need for stepping up public and private investment in infrastructure (Irrigation, rural, electrification, rural roads, warehousing and agro-processing). Secondly, increase in the total factor productivity with the use of environment friendly bio-technologies has to become a predominant source of agricultural growth. These steps reduce pressure on natural resources and can minimize the pollution from the excessive use of chemicals. Similarly, rain-fed areas need to conserve soil and moisture through watershed development programmes. Further, environment-friendly technologies in the field of animal-husbandry, horticulture and forestry sub-sectors deserve special emphasis. These tasks would require reforms at the national, state, agro-climatic regional, and panchayat level.[9-21]

Since the environment impacts of human activities are at the stage where they are not only threatening the survival of the future generations but worrying the policy-makers to impose global restrictions for future development, developing countries are required to respond quickly and positively to the rise in external environment-related interventions into their development process such as structural adjustment programmes that lead to environmental degradation, this calls for all forest products to come from sustainably managed forests. More importantly, the developing countries are increasingly required to make use of Environmental Impact Assessment (EIA) as a valuable complement to cost benefit analysis at the micro-level.[22] The environmental economics theory of sustainable development provides a framework within which to develop new planning techniques including 'hard' and 'soft' approaches to ensure environmentally safe development.

Adams (1990) pointed out that there is no reason why the present development theories cannot be modified to take account of environment and sustainable development. Development can no longer be understood apart from the environment, and since Human Development Index (HDI) emphasises human benefits, it is not an appropriate indicator of sustainable development; therefore, another index integrating the environmental and socio-economic components is needed.

## NOTES

1. E.P. Odum, (1972): *Fundamentals of Ecology*, W.B. Saunders, Philadelphia.
2. R.F. Dasmann, (1984): *Environment Conservation*, John Wiley & Sons, New York, p. 77.
3. V. Kovda *et al.* (1970): "Contemporary Scientific Concepts Relating to the Biosphere". *Use and Conservation of Biosphere*, Natural Resources' Research, X. UNESCO, Paris.
4. R.F. Dasmann, *op. cit.*, p. 115.
5. For details see: (i) FAO, *Sustainable Development and Natural Resources Management* (C 89/2), Food and Agriculture Organisation of the United, Rome, 1989. (ii) UNCN, UNEP and WWF, 'Sustainable Utilization of Species and Ecosystems', in the *World Conservation Strategy: Living Resource Conservation for Sustainable Development*, UNCN, UNEP and WWF, Switzerland, 1980.
6. R.F. Dasmaan, *op. cit.*, pp. 330-31.
7. Based on the model originally designed by Can and Underhill (1974) and later used by Hans Cristopher Reiger as a ecological model; The assumptions made under the modified version of the model in the present study are: (i) $Pt = (1+b/100)^{P}_{t-1}$. Population (P) grows at a constant rate (b) per annum over the base year (t-1) (ii) $(Fa_t) = aP_t$ (ha). To feed human population at the given technology per capita farm land (a) is needed. The total requirement of agriculture land ($Fa_t$ in year (t) is thus obtained, (iii) $L_1 + L_2$ is constant as the demand for land put to non-agricultural uses ($L_1$) and grass land as well as barren land ($L_2$) remains stationary, (iv) $F_{wt} = Z_t - E_{ft}$ Net change in the forest area ($F_{wt}$) is given by the natural growth of forest land ($Z_t$) minus the current rate of extraction of forest products ($E_{ft}$) per unit time. (v) $dF_{wt} + dF_{wt} < C_{wt}$. Increase in agricultural land ($dF_{wt}$) due to growth of population plus net change in forest area ($dF_{wt}$) takes place at the cost of available cultivable waste land ($C_{wt}$). (vi) Therefore the optimum resource capacity is reached when $C_{wt} = 0$. This model is used at micro-level in the Sutlej catchment area of Himachal Pradesh during the reference period 1981-2000 Gamrola Watershed and Kotgarh Watershed (for detail see: Analysis of Carrying capacity in selected watersheds by Amar S. Guleria in *Advances in Forestry Research in India,* Vol. III, by Ram Parkash IFS (ed.), International Book Publishers and Distributors, 1989, Dehra Dun, pp. 75-89.

8. C.H. Hanumantha Rao, (1998): 'Sustainable Development in Agriculture', *The Economic Times,* 16th Nov., p. 6.
9. Anonymous (1992): *Development and the Environment World Development Report, 1992*, Oxford University Press, Oxford.
10. Baldwin, R. (1995): "Does Sustainability Require Growth?", in I. Goldin and L.A. Winters, (ed.). *The Economics of Sustainable Development.* Cambridge University Press, Cambridge.
11. Baumol, W.J. and W.E. Dates (1988): *The Theory of Environmental Policy.* 2nd Edition. Cambridge University Press.
12. Berkes, F. and J.S. Gardner (1997): *Sustainability of Mountain Environments in India and Canada*; University of Manitoba, Canada.
13. Carrao, C. and D. Siniscalco (ed.) (1997): *New Direction in Economic Theory of the Environment*, Cambridge University Press (reprinted 1999).
14. Conard, J.M. (1999): *Resource Economics* Cambridge University Press.
15. Eckholm, E. (1975): "The Deterioration of Mountain Environments", *Science.* No. 189. pp. 764-70.
16. Harcourt,·A.P.F. (1871): *The Himalaya Districts of Kooloo, Lahoul and Spiti*, Vivek Publishing House, Delhi, (reprinted in 1972).
17. Kirk C. and M. Hobley (1993): *Socio-Economic Setting: Kullu and Mandi Districts Himachal Pradesh Forestry Project.* Vol. 11, Himachal Pradesh Government, Shimla.
18. Klemperer, W. David (1996): *Forest Resource Economics and Finance*, McGraw Hill.
19. Maxie, Jean and Baland Jean (1996): *Holding Degradation of Natural Resources: Is there a role for Rural Communities*? FAO & Clarendon Press, Oxford.
20. Sinclair, John and Laurie Ham (2000): "Household Adaptive Strategies: Shaping Livelihood Security in the Western Himalaya", *Canadian Journal of Development Studies*, Vol. 21 (i): 89-112.
21. Singh, R.B. (ed.) (1998): *Sustainable Development of Mountain Environments in India and Canada,* Oxford and IBH Publishing, New Delhi.
22. Aian Grainger (1997): "Assessing the environmental Impacts of National Development", *Sustainable Development in a Developing World,* edited by Colin Kirk Partrick and Norman Lee, Edward Eigar, Cheltenham, U.K., pp. 61-87.

# 16

# Socio-Economic and Agricultural Situation
## *A Micro-level Study in Bihar*

RAMESHWAR PRASAD SINGH

### Introduction

India is a an agrarian economy with more than two-third of the total population depending on agriculture for sustenance.* The Balthara village lies to the North-east of Basantpur Block and East of Siwan District in Bihar on the rural and agricultural area. Here, agriculture is still a dominant occupation in all respects. It is internally connected by non-metalled roads with Basantpur Block, Madarpur, Rajapatti and Nabiganj etc. (Fig. 1). The economic, social, religious and cultural life of the village is greatly influenced by the nearest towns and markets (Table 1). The main problem of the poor villagers are very low income, landlessness, unsuitable sanitary condition, etc. Famines and droughts in this area are usually due to scanty rainfall. The total geographical area of the village is 151.35 hects., population of 2751 (January 1998) uncertain with (Table 2). Most of the people are very poor. Only 38 below the poverty level. Occupational structure of population is also a pointer to that fact. The village is dominated by people working in agricultural sector, either agricultural labourers or cultivators. The village's soil is alluvial. Paddy, wheat and maize are the main staple food of the working population. More than 87 per cent of the total geographical area is under the cultivation. The percentage of irrigated areas of the village are very poor.

---

* In this paper, the socio-economic and agricultural situation has been presented as a micro-level study of Balthara village.

Due to lack of sufficient irrigational facilities, the dependence on rainfall cuts off the root of agricultural economy. Agriculture and social-forestry is the main sphere of miserable economic activities of the people. On the other hand, socio-economic conditions of the villagers vary, due to the operation of a variety of social, economic and political factors.

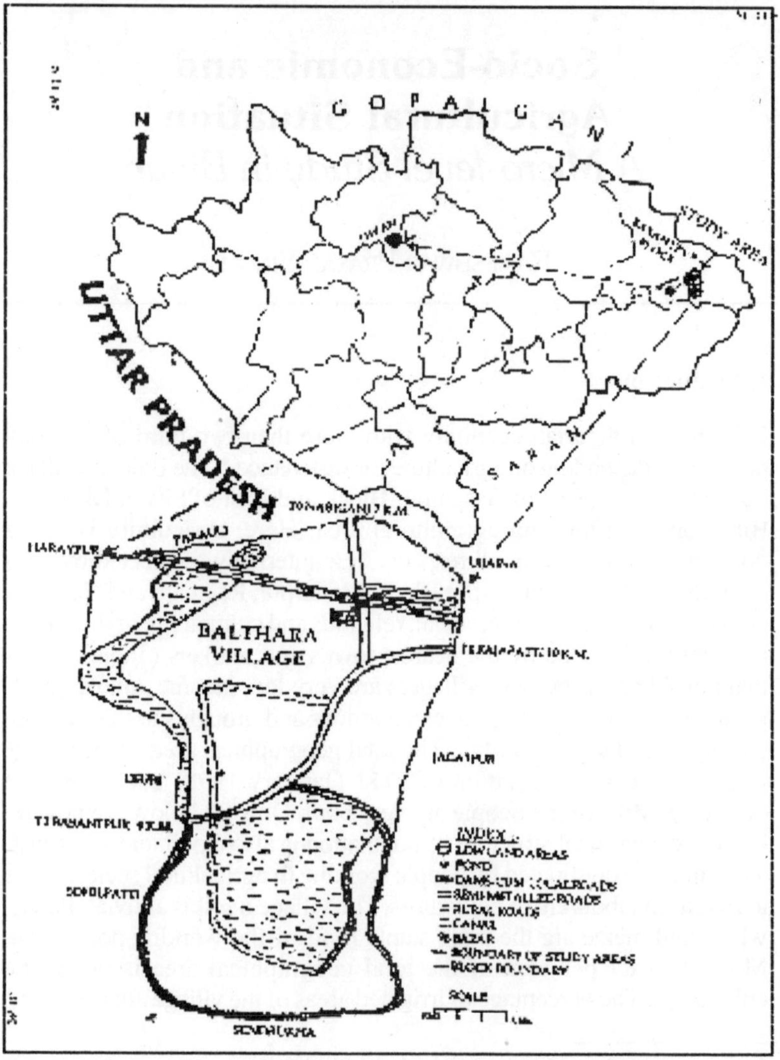


**Fig. 1:** Location Map of Balthara Village in Siwan District

**Table 1:** Income structure

| *Sl. No.* | *Income in Rs.* | *No. of Persons* |
|---|---|---|
| I | 0-500 | 225 |
| II | 501-1000 | 291 |
| III | 1001-1500 | 110 |
| IV | 1501-2000 | 57 |
| V | 2001-2500 | 21 |
| VI | 2501-3000 | 21 |
| VII | 3001-3500 | 11 |
| VIII | 351 & above | 3 |

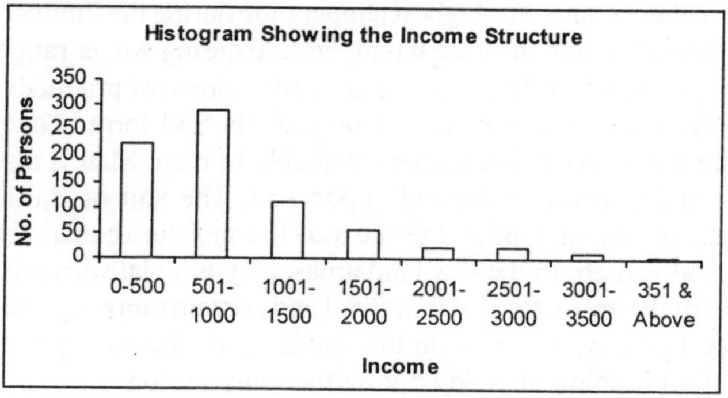


**Table 2:** Trend of population growth

| *Year* | *T.P.* | *Persons per hects.* |
|---|---|---|
| 1961 | 403 | 2.7 |
| 1971 | 729 | 4.8 |
| 1981 | 1482 | 9.8 |
| 1991 | 2017 | 13.3 |
| 1998 | 2751 | 18.2 |

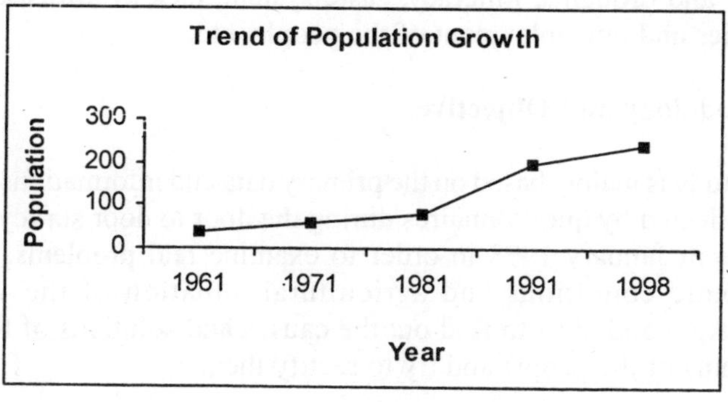

The village Balthara is situated in the Saran plain and also, presents flat plain area with gentle slopes. The village can be divided into following two natural regions: (a) Alluvial plain and (b) Low land areas, (Fig. 1). Total geographical area is 151.35 hects. and 90 per cent is agricultural land. The climatic condition of the village experiences hot-dry summers, warm-wet rains and relatively cold winters. Generally hot-dry season sets in the 1st week of April and temperature begins to rise upto June and rainy season begins from 2nd or 3rd week of June and lasts upto September. December to February is the winter season. Most of the rain occurs in July and August. Winter rainfall is very scanty and uncertain. The highest temperature during the summer ranges from 36 to 38°C and the lowest temperature during winter ranges from 13 to 17°C. Soil fertility makes a major component of productivity and therefore, is essential for better crop growth. Soil form is one of the most important natural resources available to man. Man is also child of soil and cultivation depends upon soil. The soil of Balthara, in general, are alluvial type and fertile too. The moisture contain capacity of the soil is high. In the low land areas, clay-alluvial soils are found. In most of the areas, the soil is alluvial and contain more organic matter and good quality, specially on low lands. Clay alluvial type of soil is useful for cropping of paddy during the rainy season.

The causes and motives behind rural migration are other various problems of the better part of population. One of the main causes is the insufficiency of agricultural produce in feeding the marginal and landless farmers justifying the modified dictum. The farmers are generally traditional in cultivation practice. The factors which are responsible for this great chunk of population, living low standard life are, due to old tradition and customs, low level of agricultural productivity, lack of sufficient irrigational facilities, effect of famines, floods and droughts, illiteracy, caste system, lack of infrastructural facilities and unemployment of the people, etc.

## Methodology and Objective

This study is mainly based on the primary data and informations which are collected by questionnaires during the door to door survey of the village in January 1998 in order to examine real problems, socio-economic conditions and agricultural situation of the village population and aims to find out the causes and solutions of the real problems of the people and try to rectify them.

**Problems, Status and Purpose of the Study**

The village is facing the following problems:

(i) *Low Productivity*: The socio-economic and agricultural situation of the village provides new problems to the areas which are primarily and solely agricultural and traditional in nature. The agricultural productivity of Balthara village is very low as compared to the other areas of the block. The low levels of agricultural return may be attributed to absolute technology, small size of holdings, scattered agricultural lands and inequality in ownership. Most of the farmers are tradition bound, superstitious, uneducated conservative, and their risk taking capacity is very low. They are generally reticent in adopting new technology and their poor economic condition is an impediment in the purchase of any new good quality of seeds, fertilizers, good agricultural tools and technology, development of irrigation and other infra-stractural facilities in cultivations. Paddy, wheat and maize are the staple food crops. There are two major agricultural seasons: (1) the kharif or the season of summer crops and (2) the rabi or the season of winter crops.

(ii) *Absentee Landlords*: The farmers are invariably carrying on their agriculture with the help of agricultural labourers and sharecroppers. There is a tremendous increase in the number of agricultural workers and marginal cultivators, who assist some rich farmers in the cultivation of crops. These working, marginal cultivators and sharecroppers are often exploited by the landlords. Assessment of the per capita landholding is another method to ascertain the economic condition of people. To the villagers, agricultural land is mot valuable asset.

(iii) *Unity in Diversity*: The village, agriculturally displays unity in a number of respects, such as deficiency of nitrogen in all the low productivity per unit are total cultivated areas; and per unit of labour in most areas, prevalence of backward traditional and outmoded techniques of production giving low returns, the relative scarcity of cultivated land, pre-dominance of marginal farmers, great inequality in the ownership, pattern of land production by agricultural labourers and tenants under insecure conditions, and hence risk aversion, vast seasonal and disguised unemployment, unprogressive agriculture and unsatisfactory socio-economic situation of the village. About 90 per cent of working population is engaged in agriculture, 7.8 per cent in services and 2.2 per cent in other activities. Migration of the young population is also cause of rural poverty. The old men population, poor villagers and maximum female population are also a pointer to that fact.

(iv) *Sources of Irrigation*: Water resources play a vital role in the sustenance of life on the earth surface. Groundwater may be utilised for variety of purposes including irrigation. The village agriculture is still a gamble on monsoon. The main sources of irrigation in the village are canals, wells, tube-wells, tanks and ponds. The progressive increase of the net sown area is irrigated by different sources. The percentage of irrigated areas of the village is very poor, due to, lack of sufficient irrigational facilities. Rain is the cheapest source of water providing irrigational facilities to the people. Only 21.4 per cent area is irrigated. The percentage of irrigation is TW-12.4 per cent, Canals 6.5 per cent, W-1.4 per cent and others 1.1 per cent.

(v) *Population Setting*: The distribution of population of village is controlled by physical, cultural, religious and cast factors etc. But the influence of physical aspects gradually becomes less effective with development of complex society. The ethnic, social and economic condition affect the determinants via fertility, mortality and migration of population growth. The time series analysis of population growth indicates that in 1961, the population was 43, 2.7 persons per hect., in 1971, 729, 4.8 persons per hect., in 1981, 1982, 9.8 per cent persons per hect. It was 1991, 2017, 13.3 per cent persons per hect. and in January 1998, it reached 2751, 18.2 per cent persons per hect. Literacy is a reliable index of socio-economic and cultural development. Majority of people are illiterate. The percentage of illiteracy is 72.8 per cent of this male and depending in the village. 90 per cent male migration is another major phenomenon in the village. They belong to different ethnic, religious groups and have unique social and economic characteristics. The poor people are provided with no medical facilities and any illness is liability for them. They generally, belong to scheduled castes and other backward classes. Housing occupies the most important place in the welfare works of the rural people, but housing situation of the village is very poor. The principal sources of drinking water in this area are tubewells and wells. The density of population was 9.8 per cent per hect. in 1981 and in January 1998, it reached 18.2 per cent. Of general pucka houses, 34.9 per cent are tile houses and 22.5 per cent of cottages. The growth of population of scheduled castes Muslims and other backward classes are high, due to, backwardness, illiteracy and religion, etc.

(vi) *Slums*: Slums are found almost in all countries of the world. Slums are created mainly because of poverty, social backwardness, and unemployment of the people living in the country side. The poor villagers live in slum areas, in insanitary condition. They are also affected by different kinds of diseases. About 67.4 per cent of cottages are dominated by the poor who have created the situation of slums. The scheduled castes, Muslims and other backward castes are generally residing in slums and around slums areas. These areas are dominated by poor socio-economic con-ditions in all respect. The economic and health condition is very durable in this area. The villagers generally live in badly dilapidated huts and insanitary conditions of most and the parts of village are pitiable.

(vii) *Occupational Structure*: In the village Balthara, most of the people are poor. Cultivators and agricultural labourers are the main components of agricultural workforce. Agriculture is the main occupation of the people. On the other hand, agriculture is a risky occupation of the villagers too. The village is dominated by people working in agricultural sector, either agricultural labourers, or cultivators or sharecroppers. Cultivation is the backbone of village economy and about 90 per cent of the working population are engaged therein and more than 89 per cent of total geographical areas are under the cultivation. The maximum poor people are engaged as either agricultural labourers, cultivators or daily workers. According to villager's opinion profession of agriculture work is privation due to low agricultural production in exchange of hard labour throughout the year. In January 1998, door-to-door survey reveals the fact that majority of the population are engaged in agricultural activities, i.e. cultivators, marginal cultivators, agricultural labourers sharecroppers and some other activities. At present the young male species are interested in education, service and business etc. The door-to-door survey shows that the scheduled castes and other backward castes work hard and drawing meagre amount of money. The males and females of scheduled castes and other backward classes are engaged in different types of occupation of but in case of general people, females are not engaged in any type of activity.

(viii) *Economic Activities*: Agriculture, sharecroppers, labour works and social-forestry are main sphere of economic activities of the people: Social-forestry is also backbone of the village economy and provides timber, fuelwood and fodder etc. The village economy is also dominated by agriculture and in general, the proportion of net sown area to the total area is high. Social-forestry within village area promotes not only conservation and plantation of trees but its judicial and sustainable use too. As distance between man and forest is increasing, social-forestry aims to provide employment opportunities and integrates economic gains in distribution of benefits to the village people. The histogram shows the graphical representation of the income of worker and employed persons. The door-to-door survey histogram shows that maximum number of people lie in the monthly income range of Rs. 500 to Rs. 1,000 and next to this some of the people lie in the income range of Rs. 1,000 to Rs. 1,500 and other people lie in the income above Rs. 2,000 and so. It indicates that majority of the villagers are poor and belonging to the poverty line (Table 1). The main problem of poor people is to feed their family throughout the year, very low per capita income, landlessness or possession of very little of land, very often cause starvation among them. Only 4.3 per cent of the people are of rich categories, 28.3 per cent medium categories, 33.3 per cent poor categories and 34.4 per cent very poor categories. Only 7.8 per cent of villagers are engaged in different type of simple services.

(ix) *Transport Facilities*: The rural village Balthara is located in remote backward area, where communication is very poor. Transport facilities help to make it economic for farmers to buy fertilizers and better equipments which constitutes an important item of infrastructure for socio-economic, cultural growth and agricultural development. The condition of roads in the study area is not satisfactory. Therefore, the social welfare units (schools, markets, hospitals, post office, etc.) are not well connected with the suitable roads from residential areas.

(x) *Marketing Facilities*: Rural markets are the mirrors of rural economy of developing countries. It reflects the nature of various socio-economic spectrum of the region. Marketing facilities is necessary for the village development. The accessibility to the market is a major consideration in the decision-making of the farmer. The intensity of agriculture is determined by the production of crops, declare as the location of cultivation, distance from the marketing, centres which is particularly noticeable when a bulky but low value crop has to be transported to the market. The marketing system also influences the decision making of the farmer. The suitable market facilities are not available. Balthara Bazar is only one local market, it provides basic goods to the people. The good market is located minimum 7 to 10 kms. from the village, such as Basantpur, Dighwa Dubauli; Rajapatti, Madarpur, etc.

(xi) *Capital Availability*: Capital subscribes definite limitations to the selection of crops. All agricultural inputs like livestock, irrigation, seeds, fertilizers, labour, purchase of lands, various agricultural equipments, buildings, fuel and power, maintenance require capital of the villagers make their decisions on the basis of capital to invest. The traditional way of cultivation is giving way to the market oriented crops which need more for getting higher returns. The development of village without capital is not possible, due to this maximum villagers are very poor with resultant poor economy. At present, some villagers are cultivating market and money oriented crops, such as tobacco, potatoes, oil seeds, sugarcane and vegetables. Some herdsmen also are rearing sheep, goats, pigs and poultry.

(xii) *Technology*: The technological changes including the use of modern hand-tools, animals, agricultural implements, tractors, thrashers, harvesters and more economic patterns of land management play a vital role in agricultural development. The technological availabilities of the village is very poor. They are generally traditional in cultivation practices.

(xiii) *Labours Availability*: The availability of labour is also a major constraint in the agricultural land-use and cropping patterns of the village. Labour represents all human services other than decision-making and capitalist in the village. The labour facility is of low quality and expensive too. The agricultural labourers are maximum children and women categories and their working capacities are very poor. They belong to backward classes. The rich farmers of the village are not able to manage their cultivation, due to high investment and low income.

## Conclusion

Socio-economic condition and agricultural backwardness of the villagers are clearly reflected in its inadequate social facilities. Agriculture is still a dominant occupation in all respects and cultivation is the backbone of village-economy. Most of the inhabitants are poor and this poverty is the integrating factor amongst the villages. There is a wide caste variation yet there exists a social harmony. The rate of literacy and personal income is low but new awareness has awakened the younger generation and they aspire for a self-reliant future. Famines and droughts in this areas are usually due to scanty rainfall. The farmers are generally, traditional in cultivation practices. Social-forestry is the single source of almost all the fuelwood. At present, some villagers are trying to market cash crops and also adapting animal husbandry.

## REFERENCES

Alam, S. & Siddiqi, M.F. (1984): Fertility status of soils in different Altitudinal Zones of Garwal Himalaya, *The Geographer*, Vol. XXXI, No. 1, January, p. 68.

Bagchi, A. (1996): Poverty and Wage-Labour Circulation in Two villages in Birbhum District in W.B., *Landscape Systems and Ecological Studies, ILEE*, Vol. 19, No. 1, June, pp. 20-24.

Kumar, S. (1996): Socio-Economic Study of Sahebganj Village in Raniganj P.S., W.B., *Landscape Systems and Ecological Studies, ILEE*, Vol. 19, No. 1, June, pp. 20-24.

Misra, K.K. (1996): Rural Environment of Banda District in U.P., *Landscape Systems and Ecological Studies, ILEE*, Vol. 19, No. 1, June, pp. 25-26.

Sarla, S. & Gupta, M.P. (1995): *Dynamics of Educational Development in the Chhatisgarh Region, M.P., Population Geography*, Vol. 17, Nos. 1 & 2, June-December, p. 29.

Singh, R.P. (1994): *Impact of Environmental Destruction on Tribal Population in India, East-West Geographer*, Vol. 5, No. 2, September, p. 16.

# 17

# Environmental Impact Assessment of Land Use Patterns in Kullu District of Western Himalaya

GOPAL S. SINGH

---

## Introduction

Geographical area of the Western Himalaya is about 11 per cent of the Himalayan land mass, where about 92 per cent people live in hamlets and villages. Of the total geographical area of the Himachal Pradesh, Kullu district constitutes about 10 per cent where about 93 per cent population live in rural areas and undertake agricultural activities for their prime economic mainstay. Inaccessibility, marginality, fragility, small landholdings and undulating geographical set up of the Himachal Pradesh, in general, and Kullu district, in particular, is characterised by subsistence farming practices from olden time. Hardly, fifty years back, local breed crop cultivation was the prime agricultural activity of the people. Such practices are still being maintained at the higher terraced land. Recent introduction of new on-farm technologies in the form of high-yielding varieties of food crops, fruit bearing trees and vegetables has endeavoured apparent changes in socio-cultural, socio-economical and environmental aspects. Changes in the traditional land use patterns have inevitably tempted the farmers to mould their tendencies to the market based resources and forces. This *inter alia* includes changes in the traditional farming practices, increasing the competition amongst the available human and natural resources, uncontrolled use of yield increasing inputs like chemical fertilizers, pesticides, insecticides and fungicides massive loss of tree cover at the cost of introduction of fruit bearing trees, decreasing pace of heterogeneity of crops (loss of crop biodiversity), increasing dependency of farmers on the market based resources and availability which have altogether accelerated the environmental damages compounding integrated effect on farmers health and socio-economic condition of the subsistence farmers. The quantum of use of chemicals as pesticides, insecticides and fungicides has increased at an alarming rate parallel to the apple tree introduction in the Kullu district. If the reckless use of chemicals is not checked and the doses recommended and popularized based on the scientific study then one could not be able to set forth the future of the traditional system, environmental changes and health status of the users.

Three villages, namely Jigala in Banjar valley, Jhiri and Jater in Kullu valley were selected in Kullu district (Jhiri is the ecotonal boundary of Kullu and Mandi district but falls in Kullu valley) after the primary reconnaissance of the area. These villages are representative enough to represent all characteristics of the Kullu district.

Kullu district is located in the heart of the state of Himachal Pradesh in the North-western Himalaya. It is surrounded by Lahaul and Spiti district from north and east, by Shimla and Kinnaur districts from south and south-east direction, by Kangra and Mandi districts from west and south-west and covers geographical area of about 5500 sq. km. In general, Kullu valley has difficult terrain with gorges, ravines and extreme climatic variability. The valley is popularly known as *Dev Bhumi* (land of God). The valley spread on either side of river Beas, covering a distance of about 85 km. between Aut at the lower part and Rahla at the upper part of the valley. The width of the valley varies from a few metre to about 3 km. supporting varieties of land use types. The valley is surrounded by the higher hill reaches in a continuous chain. The far-flung hamlets are located across the valley commencing from hill base closed to river beds and spread on middle terraces. On the basis of the micro-climate variability, the valley has been divided into two parts, viz., the lower Kullu valley (area from Aut to Kullu less than 1200 m) upper Kullu valley (area from Kullu to Manali more than 1200 m). Village Jhiri (900 m) is located in lower zone and Jater (1400 m) is located in the upper part of the Kullu valley. The third village—Jigala is located in Banjar valley at an elevation of about 2100 m. The Banjar valley lies along the Tirthan river. This valley is not as wide as Kullu valley. The introduction of fruit trees particularly at higher elevation in this area is recent phenomenon.

Hindus constitute the majority of population, about 96 per cent of whom are located in Kullu district. The people and the culture of the Kullu valley are specifically identified by the collective term, pahari (of the mountains). The community is dominated by Hindu, Buddhist and Islamic culture. Nonetheless, blend culture of the people of the Kullu valley closely follows the distinction among local people, people migrated from the plains, Nepalis, Tibetans and Kashmiris. From isolated hamlets to conglomerated settlements are the characteristic features of the villages in the Kullu valley.

Salient features of the study villages are given in Table 1. Village—Jhiri and Jater are located in Kullu valley, former falls in the lower portion of the valley and latter in the upper portion of the valley. Some of the infrastructural facilities were also noted in these villages. Livestock diversity is considerably higher in Jigala, the village located at higher altitude as compared to villages located at the lower altitude in the valley. This indicated that livestock play a catalytic role in upgrading the village economy of the poor and marginal farmers who are located in the remote areas. Apple introduction in such higher located villages is relatively a new phenomenon. Fruit bearing trees diversity is invariably higher in the case of valley land as compared to terraced land. Fruit producing trees particularly apple trees were initially introduced in the lower parts of the valley but now have gradually started shifting to middle and higher located terraced land. Owing to unprofitable nature of apple produce in the valley land, farmers have recently started growing vegetable crops and earning exorbitant amount of cash.

**Table 1:** Salient Features of the Study Villages at a Glance.

| *Parameters* | *Jigala* | *Jhiri* | *Jater* |
|---|---|---|---|
| **General Features** | | | |
| Name of village and Panchayat | Jigala and Palach | Jhiri and Jhiri | Jater and Katrain |
| Location in valley | Banjar | Kullu | Kullu |
| Name of river | Tirthan | Beas | Beas |
| Distance from Kullu district headquarters (approx. km.) | 35 | 15 | 20 |
| Elevation of village (m) | 2100 | 900 | 1400 |

(*Contd.*)

*(Contd.)*

| *Parameters* | *Jigala* | *Jhiri* | *Jater* |
|---|---|---|---|
| Health centre | No | No | No |
| School (class) | Middle | No | Primary |
| Electricity | Yes | Yes | Yes |
| Post office | Yes | No | No |
| Motorable road | No | Yes | Yes |
| Small shop | Yes | No | Yes |
| Ration supply shop | Yes | Yes | Yes |
| Drinking water facility | Poor | Poor | Poor |
| Use of chemicals | Yes | Yes | Yes |
| **Socio-economic profile** | | | |
| Total No. of households (aprox.) | 25 | 125 | 70 |
| Total population (approx.) | 225 | 750 | 550 |
| Diversity of livestock | High | Low | Medium |
| Dominant crops | Wheat | Vegetable | Wheat |
| | Maize | Wheat | Maize |
| | | Rice | Rice |
| | Apple | Apple | Apple |
| | Plum | Plum | Plum |
| | | Almond | Pear |
| | | Pear | Almond |

Table 2 shows the composition and the demographic profile of the study villages. Total landholdings in village located at the higher altitude is considerably lower as compared to valley located village. The higher human population of the upland located village could significantly be correlated with the remoteness of the area which needs relatively more human power for their agricultural operations. Similarly, number and diversity of livestock is significantly more in the upland village which shows that rearing of livestock subsidises the economy of domestic and agricultural sector. The sex ratio (females per 1000 males) is remarkably higher in the study villages (1048-1133) as compared to the district level (1020) and state level (975). The percentage of total literacy (26.7% to 46.6%) in the surveyed village is sharply lower than that of Kullu district (54.8%) and the state of Himachal Pradesh (63.9%). Male literacy is relatively higher as compared to female literacy. Lower literacy rate in the study villages could be due to lack of awareness of education and educational facilities in conjugation with their lower priorities for education. Agriculture is the prime occupation of the village people. It has been found that about 96 per cent of the workforce were engaged in cultivation and remainings were engaged outside the village boundary either in some wage paid activities or in some business activities. Of the total cultivators about 20 per cent work as labourers on daily wages in or around the village whenever they are free from their own agriculture work.

**Table 2:** Structure and Demographic Profile of the Study Villages.

| *Parameters* | *Jigala* | *Jhiri* | *Jater* |
|---|---|---|---|
| Total households surveyed (No.) | 15 | 15 | 15 |
| Total cultivated land (ha) | 22 | 33 | 27 |
| Population (No.) | 101 | 96 | 86 |
| Male | 48 | 45 | 42 |
| Female | 53 | 51 | 44 |
| Children | | | |
| 0-5 | 1 | 19 | 9 |
| 6-1 | 17 | 15 | 13 |
| 10-15 | 15 | 15 | 15 |
| Sex ratio | 1104 | 1133 | 1048 |
| Livestock population (No.) | 176 | 59 | 44 |
| Goat | 3 | 2 | – |
| Sheep | 106 | 35 | – |
| Cattle | 67 | 22 | 44 |
| Literacy (%) | | | |
| Male | 40.0 | 60.0 | 53.3 |
| Female | 20.0 | 33.3 | 33.3 |
| **Total** | **26.7** | **40.0** | **46.6** |

## Socio-economic Profile

Socio-economic attributes of the study villages at households level have been illustrated in Table 3. The family size varies from 5.7 in the upper portion of the Kullu valley (Jater village) to 6.4 in the lower part of the valley (Jhiri village). However, 6.7 family size is noted in the case of Jigala village which is located at the higher elevation. Similarly, per capita land availability is remarkably lower in up located settlement as compared to valley land featuring the availability of flat land in the valley. Sufficient amount of availability of fodder and free grazing land closed to forest fringe allows farmers to rear varieties of animals at the up located village which not only helps in upgrading the village economy but also subsidises the agriculture and domestic sector. Unavailability of free grazing lands coupled with meagre production of fodder due to fast spreading horticulture and agriculture in valley in valley land has curtailed the rearing of livestock. From fields having an age of greater than 20 years fruit trees do not favour for the food crop cultivation because of shade effect imposed by these trees. High fruit crop diversities and net areas under plantation were recorded in the lower valley land as compared to the upper located terraced land. High cropping intensity in up located village land invariably favours for better agronomic yield. Contrary to this, fruit production is remarkably higher in the valley land due to its intense diversity and maximal land area cover under fruit trees. Unsuitable environmental condition and spreading tree diseases have curtailed the apple cultivation particularly in the lower portion of Kullu valley and people have recently started cultivating varieties of vegetable crops. Farmers have started cutting old apple trees which have become unprofitable and uneconomical in nature.

**Table 3:** Socio-economic Attributes (Per family) of the Study Village.

| *Parameters* | *Jigala* | *Jhiri* | *Jater* |
|---|---|---|---|
| Total agriculture land (ha) | 1.47 | 2.20 | 1.80 |
| Family size | 6.73 | 6.40 | 5.73 |
| Availability of per capita land (ha) | 0.22 | 0.34 | 0.31 |
| Cattle (No.) | 4.47 | 3.80 | 2.93 |
| Sheep and goat (No.) | 7.28 | 0.13 | – |
| Net cropped area (ha) | 1.58 | 1.14 | 0.51 |
| Wheat | 0.70 | 0.30 | 0.18 |
| Maize | 0.67 | 0.42 | 0.17 |
| Miscellaneous | 0.19 | 0.13 | 0.11 |
| Vegetables | 0.02 | 0.29 | 0.05 |
| Net fruit crop plantation (ha) | 0.50 | 1.58 | 0.99 |
| Apple | 0.30 | 1.08 | 0.79 |
| Plum | 0.20 | 0.24 | 0.13 |
| Pear | – | 0.17 | 0.07 |
| Almond | – | 0.07 | – |
| Gross agronomic yield (kg.) | 2600 | 1655 | 1200 |
| Food crop purchased from market (kg.) | 120 | 550 | 770 |
| Fruit yield (kg.) | 1100 | 3200 | 3550 |
| Monetary gain from fruit trees (Rs.) | 12000 | 28000 | 30250 |
| Monetary gain from vegetable crops (Rs.) | – | 12000 | 225 |

## Usage of Agro-chemicals

Intensity of chemicals (pesticides, insecticides and fungicides) use for the apple trees is sharply high in Kullu valley, as 87-93 per cent farmers have reported (Table 4) usage of chemicals. Use of chemicals for vegetable crops is drastically higher at the lower portion of the valley as compared to upper terraced sloppy land. Well geographical feasibility and adequate water availability in conjugation with the suitable market facility and demands are the prime causes of accelerating vegetable cultivation at the lower portion of Kullu valley. Farmers harvested cabbage thrice and tomato, brinjal and chilli twice from a particular field in a year while apple is harvested only once in a year.

**Table 4:** Per cent Response for the Use of Chemicals

| *Name of crops* | *Jigala* | *Jhiri* | *Jater* |
|---|---|---|---|
| Fruit trees | | | |
| Apple | 80.0 | 93.3 | 86.7 |
| Plum | 20.0 | 53.3 | 33.3 |
| Pear | – | – | – |
| Almond | – | – | – |
| Vegetables | – | 80 | 40 |
| Food crops | – | – | – |

*Source*: After Singh, 1996.

Farmers have been cultivating variety of apple trees in the area to ensure optimal fruit production (Table 5). Intensification of apple cultivation has resulted massive use of non-renewable resources like pesticides, insecticides, fungicides and chemical fertilizers. Market demand and subsidy provided by the Government agency, by and large, have accelerated the usage of chemicals in the agriculture fields. Because of the use of heavy dose of harmful chemicals, multitude of diseases on the trees and fruits have increased at an alarming rate of course, highly market demanded apple trees are more prone to the infection of pests, insects and fungus which have resulted in massive use of pesticides, insecticides and fungicides. With the increasing demand of the coloured apple, varieties of chemicals have categorically been used in the area which are merely responsible for colouring of apple, as responded by farmers/growers. Some of the common diseases of apple trees are canker (peelings of trees bark and finally trees dry), scale (red patches on trees trunk allowing no growth of tree) and scab (small red spots which not the fruit).

**Table 5:** Intensity of Chemicals Used for the Different Variety of Apple Trees and Vegetables being Grown in the Study Area

| *Apple Varieties* | *Chemical Use* | *Market Value* |
|---|---|---|
| Commercial | High | High |
| Early royal | High | High |
| Golden | Low | Low |
| Green | Low | Low |
| Jonathan | Low | Low |
| Kalidevi | Low | Low |
| Red | Low | Medium |
| Red golden | Medium | High |
| Red june | High | Medium |
| Rika red | Medium | Low |
| Royal | High | High |
| Vinter | High | Low |
| **Vegetables** | | |
| Cabbage | High | High |
| Tomato | High | High |
| Brinjal | Medium | High |
| Chilli | Low | High |

*Source*: After Singh, 1996

## Recommended Chemicals for Apple Trees Use

Table 6 shows the list of chemicals in the form of insecticides, pesticides and fungicides as recommended by the horticulture department for spraying. Minimal eight time spraying at the different stage of apple trees starting from sprouting to flowering, fruiting and even after harvesting of apple has been prescribed by the horticulture authorities. Depending upon the importance of disease and money availability to the subsistence growers, some of the chemicals like Dodine and Dithenon are even sprayed twice and thrice in one crop season. In spite of the recommendation and due to high cost of the chemicals, poor farmers hardly sprayed 3-4 effective chemicals like Dodine, Dithianon, Captap and Fultop. This table also throws some light pertaining to the amount of chemicals to be used in the form of spraying but it does not show neither the number of trees to be sprayed nor age of the plants. Marginal farmers do not have even exact measurement for making the mixture of water and chemical; they make the mixture arbitrarily. Some of the chemicals which are not in the list of horticulture department are being frequently used by the farmers. Synthetic growth regulating chemicals like Ethfol (responsible for colouring of apple) and multiplex (making attractive size of apple) which are not recommended by the Horticulture department are frequently being used by the growers.

**Table 6:** List of Chemicals Recommended by the Horticultural Department (Chemical per 100 litres of water)

| | *Trees stage* | *Chemicals* |
|---|---|---|
| 1. | Silver tip to green tip | Chlorothalonil (400g)/Dodine (100g)/Mancozeb (400g)/Dithianon (75g)/Captan (300g) |
| 2. | Pink bud stage | Bitertanol (50g)/Fenarimol (40ml)/Myclobutanil (40g)/Mancozeb (300g)+Sulphur (200g)+Folithion (100ml)/Dursban (100ml)/Danusban (100ml)/Durmet 20EC (100ml) |
| 3. | Petal fall | Carbendazim (50g)/Thiophanate methyl (50g) |
| 4. | Fruit set (pea size) | Dodine (75g)/Bitertanol (50g)/Fenarimol (40ml)/Myclobutanil (40g)/Mancozeb (300g)/Captan (300g) |
| 5. | Fruit development (walnut size) | Carbendazim (25g)/Thiophanate methyl (25g)+Mancozen (250g)/Captan (300g)/Dithianon (50g)+Metasystox 25EC (100ml)/Rogor 30EC (100ml)/Accothion 50EC (100ml)/Dimecron (35ml)/Dursban (10ml)/Danusban (100ml)/Durmet 20EC (100ml) |
| 6. | Fruit development | Mancozeb (300g)/Captan (300g)/Bitertanol (50g)/Fenarimol (40ml)/Myclobutanil (40g) |
| 7. | Pre. harvest | Mancozeb (300g)/Captan (300g)+Carbaryl (Sevin 50 WP) (100g)/Malathion (Cythion 50 EC) (100ml) |
| 8. | Post harvest spray | Fenitrothion (Sumithion/Folithion/Accothion 50 EC 100ml)/Chlorpyriphos (Dursban/Danuban/Durmet 20 EC 100ml) |
| | *Recommended brand names of fungicides*: | |
| | Carbendazim | = Bavistin/Derosal/JK Stein/Bengard/Agrozim/Kilez/Dhanustein; |
| | Mancozeb | = Indofil M-45/Dhanutan/Bengtan/Hexacap/Kohicap; |
| | Dithianon | = Delan; |
| | Thiphante methyl | = Topsin M; |
| | Bitertanol | = Baycor; |
| | Chlorothalonil | = Kavach; |
| | Dodine | = Syllit/Scabicide/Superstar; |
| | Fenarimol | = Rubigan; |
| | Myclobutanil | = Systane |

*Source*: After Singh, 1996.

Apart from the apple trees varieties of dangerous chemicals being used for vegetables. Table 7 shows the list of insecticides being used by the vegetable growers. Insecticide like Nuvon is highly dangerous and this chemical is very common for suicidal trial. These chemicals are easily available at private shop. Nuvon is also used for killing the flies, pissu and rat. Similarly, zink phosphate is used for killing of rat. Cypermethrion is used for flies killing. Aluminium phosphate is used for storing of wheat grains which is also dangerous.

**Table 7:** Chemicals Used for Vegetables in 100 Litre Water

| *Chemical* | *Quantum* |
|---|---|
| Indocelphon | 200g |
| Cypermetherion | 100ml |
| Fenbelret | 100ml |
| Nuvon | 75ml |
| Monocrotofos | 100ml |
| Malathene | 200ml |

*Source*: After Singh, 1996.

Table 8 reveals the list of chemicals which have been distributed by the Department of Horticulture, Kullu district, covering three blocks, namely Nagar, Kullu and Banjar. Quantum of distribution has tremendously decreased even after increasing area of apple plantation. Fultop used for colouring and making bigger size of apple has now been banned by the horticultural department in 1994-95 but, still being most preferred chemical by the farmers and easily available in the market. This chemical is suppose to be highly inhuman for the health perspectives.

## Awareness of Chemical to the Farmers

Table 9 depicts about the awareness of chemicals to the users group. It has been noted that hardly 27-40 per cent farmers use the recommended dose of chemicals as prescribed by the Horticulture Department (cf. Table 6). Majority of the farmers (67 to 100%) purchased chemicals from privately owned shop. Government at the time of peak demand of the growers and need of the plants, as claimed by the majority of the farmers, fails in supplying chemicals. Farmers having higher landholdings and better economic position are the prime beneficiaries of the subsidy.

**Table 8:** Amount (kg.) of Chemicals Distribution from District Horticulture Office.

| *Chemical* | *1990-91* | *1995-96* |
|---|---|---|
| Dodine | 6407 | 2557.2 |
| Diathane M45 | 43715.5 | – |
| Fultop | 1399.8 | – |
| Bakore | – | 569.9 |
| Mancozeb | 587 | – |
| Captan | 1962 | 10835.5 |
| Vangard | 87.2 | – |
| Topsin M | 1710.5 | 464.6 |
| J.K. Stein | 617.8 | 58.5 |
| Agrojim | 213 | 1149.5 |
| Carbendazin/Bavistin | 801.6 | 36.6 |
| B Stene | 411.6 | – |
| Mancozeb | 5428.5 | – |
| Dhanuka M45 | – | 1752 |
| Indofil M | – | 29528 |
| Rubigen | – | 702 |
| **Total** | **63341.5** | **47654.2** |

Despite of the subsidy provided by the Government machinery (50% subsidy for marginal and small farmers and 30 per cent subsidy for larger farmers), small and marginal farmers could not avail the benefit of subsidy because of lack of coordination. Farmers are even not aware of using mask and goggles at the time of spraying. Very few users (27-53%) have responded washing of cloth after spray and hardly 60 per cent sprayers took bath after the spray. Eating and packing of apple without washing is common amongst the villagers. Use of empty chemical bottles and packages (which are usually bigger in size) is common just after mild washing. Packing of empty packets (which think to be of no use) in deep soil is surfaced, as reported by 7-20 per cent farmers. Farmers claimed that most of the pesticides are not effective. None of the technical know-how person from the Government office has visited to these three villages. Due to lack of facilities and knowledge, exact mixing of chemicals with the water is not known to the users. Therefore, there is strong need to provide training and basic facilities to the farmers in this regards.

**Table 9:** Farmers' Responsc (%) of Awareness of Chemicals Use and Precautions While Using.

| *Parameters* | *Jigala* | *Jhiri* | *Jater* |
|---|---|---|---|
| Use of recommended dose | – | 26.7 | 40.0 |
| Purchased from privately owned shop | 100.0 | 66.6 | 80.0 |
| Purchased from government recognised shop | – | 34.4 | 20.0 |
| Recommendation of use of chemicals by privately owned shop | 100.0 | 73.3 | 86.7 |
| Recommendation of use of chemicals from government recognised shop | – | 26.7 | 13.4 |
| Use of mask | – | – | – |
| Use of goggles | – | – | – |
| Washing of cloths after spray | 26.7 | 46.7 | 53.3 |
| Taking bath after spray | 20.0 | 60.0 | 60.0 |
| Washing of apple before eating | – | – | – |
| Washing of apple before packing | – | – | – |
| Re-use of packets/bottles | 60.0 | 73.3 | 60.0 |
| Do you buried empty chemical packets/bottles | 13.3 | 6.7 | 20.0 |
| Ineffective chemicals | 53.3 | 40.0 | 40.0 |
| Visiting of officer | – | – | – |
| Exact mixing of prescribed chemicals | 6.7 | 20.0 | 20.0 |

## Health Problems

Table 10 summarises the health problems faced by the chemical users. Spraying is mostly performed by the younger members of the family with the collaboration and support of the neighbouring families. Some of the large landholders engaged Nepalis on wage basis. Majority of the chemical users feel headache, giddiness, etc. at the time of spraying. Skin problems, respiratory difficulty, eyes infection and dysentery problems are too common. Some people perceived that dysentery problem has increased tremendously particularly in rainy season. This could be because of the contamination of ground water with the varieties of chemicals. Suicidal trial which has been never heard by the isolated hill society is now being common to hear. This could be definitely correlated with the introduction of the insecticides and pesticides in the area.

**Table 10:** Farmers Response (%) on Health Imbalances/Problems Caused by Chemicals

| *Parameters* | *Jigala* | *Jhiri* | *Jater* |
|---|---|---|---|
| Headache and giddiness | 94.6 | 100.0 | 94.6 |
| Vomiting | 20.3 | 13.3 | 13.4 |
| Stomach pain | 46.6 | 40.0 | 40.0 |
| Dysentery | 52.6 | 40.0 | 46.6 |
| Skin allergy | 60.0 | 46.7 | 53.3 |
| Eye problem | 40.0 | 26.7 | 40.0 |
| Respiratory/breathing difficulty | 46.6 | 73.3 | 60.0 |
| Weakness | 80.0 | 93.3 | 100.0 |
| Body pain | 80 | 100.0 | 93.3 |
| Sucidal attempt* | 13.3 | 40.0 | 46.6 |

* Around the village.

*Source*: After Singh, 1996.

Introduction of chemicals is much older in the lower parts of the valley as compared to the upper parts; introduction at higher elevations is very recent phenomenon (Table 11). Application of insecticides and pesticides on the vegetables is hardly 4-6 years old. Intensity of use of chemicals (pesticides, insecticides and fungicides) has increased tremendously since last 20-25 years once after the massive introduction of the apple trees in the area as initiated by Government agency on heavy subsidy. About 93 per cent farmers have responded that varieties of disease have recently invaded in the area which have led an evolution of use of heavy dose of pesticides, insecticides and fungicides. Particularly in the lower part of the Kullu valley where dose and intensity of chemicals use is invariably high, has attracted and invited varieties of diseases in the area. People have responded that varieties of new born human diseases have come up in the area. Local people feel that intensity and frequency of dysentery problems in the area has increased tremendously and this could be because of contamination of ground water with the chemicals. Such problems are more common in rainy season and majority of the people feel that it could be a new type of disease in the area. Similarly, stomach pain is a common claim of the majority of the farmers and they strongly feel that it is a new born disease and could be because of the more chemical use in the area.

**Table 11:** Possible Introduction of Diseases in the Area.

| *Parameters* | *Jigala* | *Jhiri* | *Jater* |
|---|---|---|---|
| Introduction of fruit trees | 1980s | 1960s | 1970s |
| Invasion of new born disease on human beings | 13.33 | 66.7 | 46.7 |
| Disease on fruit trees | 40.0 | 93.3 | 86.6 |
| Disease on vegetables | – | 86.7 | 13.3 |
| Dysentery | 6.67 | 46.7 | 40.0 |
| Stomach problems | 6.67 | 26.7 | 33.3 |

## General Health Status

Good health status of the people might have co-evolved with the existing difficult geographical terrain and prevailing harsh environmental condition tuned to the area. In general, health profile of the people are good. However, old people feel that quality of health status is deteriorating gradually because of increasing impurity of food quality. Per capita availability of milk has gone down due to ever-increasing population pressure and decreasing the cattle husbandry. Medicine is easily available because of availability of facilities. People settled at higher elevation believe in traditional medicine which is easily available in endowed forest. Fatal disease like tuberculosis was noted in the study villages. Some of the people feel that changing landuse pattern is faltering the health status of the people.

## Changing Traditional Land use System

Kullu valley was once popularly known for subsistence farming operation tuned to the region. Large number of people were cultivating local breed staple food crops but such trends have reversed due to massive introduction of the high yielding variety of food crops and fruit bearing crops in the area; introduction of vegetable cultivation is a very recent phenomenon. Massive introduction of fruit trees has led an evolution of heavy use of non-renewable resources in the area leaving aside the traditional knowledge based land use operations. Changing landuse pattern is root cause of commercialisation of agriculture in the area. Fast introduction of apple cultivation has imposed adverse impact on natural as well as man made systems. Overall ecological and environmental changes were noted in the area due to apple introduction. Not only crop diversity and cropping pattern has altered but pressure on the natural system in form of fodder, fuelwood and timber has increased many fold. Changing land use scenario has changed the overall monetary profit of the farmers but it has paralysed the traditional knowledge of the area. Invasion of chemical fertilizers, pesticides, insecticides and fungicides in the area is hardly 20-25 years old and is merely resulted of the introduction of new on-farm technologies in the form of high yielding varieties of fruit crops.

**Strategies and Priorities for the Sustainability**

It has been found that traditional agricultural practices that have evolved locally and have been inheriting since generation are now gradually being replaced with the apple introduction in the area. Heavy subsidy provided by government agency on different items is the root cause of accelerating apple cultivation in the area. Subsidy should be continued provided marginal and small farmers would crop the maximal benefit. There is strong need to study the different land use system and subsequently to formulate the strategies for sustainability of the systems. However, developmental priorities and strategies could be centred around agricultural based activities which would improve the quality of life of the small and marginal farmers. Lack of technical know-how due to meagre training support from the skilled department could be the sole cause of improper use of varieties of dangerous chemicals. Proper training to the farmers through the Horticulture Department could narrow down the gap incurred in the knowledge and traditional system. Poor farmers are not the real beneficiaries of the subsidy as provided by government agencies. Proper and equal distribution of on-farm wealth could accelerate and subsidise the economy of the marginal farmers. Tendency of copying the use of pesticides, insecticides and fungicides by the subsistence farmers would possibly be developing varieties of resistance pests and insects. There is strong need to design, develop and test bio-chemicals in form of bio-pesticides and bio-insecticides. Recommended minimal eight spray could be curtailed down to 2-3 based on scientific study. Integrated training of the growers with the help of government machinery, non-government institution and planners in light of on-farm management could upgrade the knowledge and economic conditions of the farmers.

**Major Findings and Recommendations**

- In the wake of change of traditional land use patterns, the use of new on-farm technologies in the forms of horticulture tree species and non-renewable resources have increased tremendously.
- Introduction of chemicals in the forms of insecticides, pesticides and fungicides in the study villages has been started hardly 20-25 years back; the pace has increase tremendously over the past few years due to the apple introduction in the farm field.
- Proper use of the dangerous chemicals is not well known to the subsistence farmers.
- Some of the banned dangerous and harmful chemicals are frequently being used by the farmers.
- Availability of desired chemicals at required time to the farmers is lacking from the Government agency.
- Marginal farmers are not properly getting the benefit of subsidy as provided by Government agency.
- Varieties of chemicals usage have subsequently resulted a number of short-term and few long-term health impacts on human beings; intensity and frequency of new born diseases have increased in the area.
- Overall ecological, environmental and socio-economic changes have been noted in the study villages.
- Developmental priorities should be centred around agri-cultural based activities.
- Integrated training of chemicals users with the help of Government agency, planners and scientific support would promote technical know-how and socio-economic status of the farmers.
- Number of recommended chemicals in the form of insec-ticides, pesticides, and fungicides should be curtailed down to minimal number based on the proper scientific study and introducing bio-chemicals.
- Infrastructural facilities like mask; goggles and body covers would be provided on subsidized price.
- More technological alternatives are still needed. Knowledge is sorely lacking about sustainable on-farm management. This might fulfil through integrated training.

- High-level policy-makers, researchers and experts from non-government organization would work together to find out the better planning in terms of alternative use of harmful chemicals.

**Acknowledgements**

The author is grateful to Society for Participatory Research in Asia, New Delhi, for providing financial support and Mr. H. Jaitli, for providing basic facilities and in publishing this report in Occupational and Environmental Health.

# 18

# Spatial Pattern of Carrying Capacity of Land in Bilaspur District

KAVERI DABHADKER

Agriculture plays a vital role in the economy of the rural region. It is imperative to analyse the arable land resource availability and future need of the population. In Bilaspur district of Chhattisgarh where 89.8 per cent of population is engaged in agricultural activities, with moderate population growth, there is a need for measurement of its carrying capacity of land to get the answer of the questions, i.e. whether the available land is productive enough to meet the requirements. Although, the productivity can be increased to some extent, yet it cannot feed the indefinite number of people. It has its limitations depending upon the quality of land, types of land, utilization and lands technologies used (Reddy & Reddy, 1993). A study of carrying capacity of land of a backward district, will better reveal the man-land relationship and would perhaps throw some right on the sustainability of the existing land resources.

Carrying capacity of land is defined as the number of people that a unit area of land can support at a given technological level and system of land use (Gatade, 2004). The concept of carrying capacity was introduced by Dudley Stamp. According to Stamp (1958), the carrying capacity is a measure of farming efficiency and it is calculated by converting crop production into caloric values.

## Study Area

Bilaspur district is located in the northern-eastern part of newly formed state Chhattisgarh between 81°21′ east to 21°41′ west and

23°70′ north latitude, over an area of 857 sq. km. Physically, it can be divided into two parts: northern hills and plateaus and southern plain (240 metres above sea mean level). District is drained by Arpa, Maniyari, Agar rivers, the tributaries of Sheonath river system. Soil is red yellow and if one looks at the land use, 43.5 per cent is under cultivation and about 13.19 per cent of land is covered with forest. Paddy is the main crop grown in 63.28 per cent of total cropped area. Total population of the region is 1,99,642 (2001) with annual growth rate of 17.59. Average density of population is 232 persons per sq. km.

## Objective

The main objective of the present study is to calculate the carrying capacity of crop land in Bilaspur district based on the standard nutrition unit for production.

## Data Base and Methodology

This study is based on secondary data for food crops, taken from Directorate of Agriculture, Bilaspur, for the year 2004-05. The carrying capacity is measured in terms of caloric output of paddy crop, which occupied more than 95 per cent of the net sown area in the study region, as rice is the staple diet.

Paddy is the dominant crop of the study area, so the data for paddy crop and average crop yield of paddy for the year 2004-05 is taken into consideration.

The production of paddy crop is multiplied by its calorific value and after the deduction of 20 per cent as wastage of total production actual caloric value is calculated. To obtain the per hectare food crop in calorie Potential Production Unit (PPU) following formula is used:

$$\text{PPU (Calories/hectare/year in million)} = \frac{\text{Calorie value of the paddy crop}}{\text{Area under paddy crop}}$$

By using Standard Nutrition unit for production (suggested by ICMR, Hyderabad is adopted which is 1,051,200 Calories/person/annum) and the PPU, the carrying capacity in terms of person/hectare/year is calculated for each tehsil, as given below:

$$\text{Carrying Capacity} = \frac{\text{PPU (Potential Production Unit)}}{\text{SNU (Standard Nutrition Unit for Production)}}$$

For the study of pressure on land, food-crop density (per head land availability) is also calculated using the following parameter:

$$\text{Food crop density} = \frac{\text{Population of an areal unit}}{\text{Area under paddy crop in an areal unit (food-crop density)}}$$

After calculating Tehsilwise per hectare availability of land, it was compared with the carrying capacity of the land. Based on this comparison, surplus/deficiet area of food identified.

### Distribution Pattern of Potential Production Unit (PPU)

The PPU is expressed in terms of Caloric productivity/hectare/annum in the region. In the period of 2004-05, the district had a caloric productivity of 77.1 million calories/hectare/year. The PPU ranged spatially from as low as 59.1 million/hectare/year in Pendra to 99.2 million/hectare/year in Lormi. Three categories emerged on the basis of the distribution of PPU:

1. High PPY region (80 above million calorie/hectare/year).
2. Medium PPU region (60-80 million calorie/hectare/year).
3. Low PPU region (4-60 million calorie/hectare/year).

### High PPU Region

This category included Lormi, Takhatpur and Mungeli, Masturi Tehsil of the study area. In this region, due to availability of irrigation facility through Kharang, Khudiya medium irrigation project, Tehsil-wise double-cropped and single-cropped area should be given in a Table, double cropped area increased and provided better situation for agricultural innovations to achieve higher production (Fig. 1).

### Moderate PPU Region

Bilha and Kota Tehsil, belong to this category. Unavailability of surface and limited ground water as well as undulating physical profile specially in northern Kota Tehsil restricted the agriculture practice.

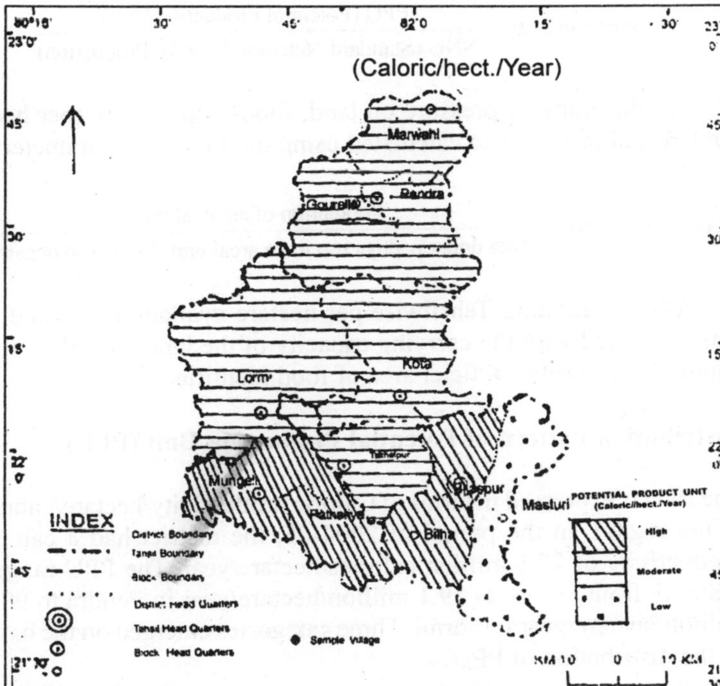


*Source*: Office of Land Record, Bilaspur. Low PPU Region.

**Fig. 1:** Bilaspur District: Distribution of P.P.U.

Northern hilly and forested Pendra tehsil represents this category. Productivity of this region is low because of rugged terrain and predominance of Bhata and gravelly soils. Other causes of low production are low use of fertilizer, old type of agricultural operation, small holdings, etc.

## Spatial Pattern of Carrying Capacity

The carrying capacity of land is expressed in terms as—SNU and PPU which gives the number of people, a unit land can support per year.

The average carrying capacity of Bilaspur district is 7.7/person/hectare/annum. It varied from as low as 5.6 persons/hectare/annum at Pendra to as high as 9.4 persons/hectare/annum in Lormi tehsil.

Spatial distribution of carrying capacity is not much different from PPU distribution. In northern hilly and forested rugged terrain had very low carrying capacity, while in southern part of the district, carrying capacity is comparatively high. On the basis of carrying capacity the tehsils may be divided into four categories:

1. Low carrying capacity area: <5 persons/hectare/year
2. Moderate carrying capacity area: 5/7 persons/hectare/year
3. High carrying capacity area: 7/9 persons/hectare/year
4. Very high carrying capacity area: >9 persons/hectare/year

**Low Carrying Capacity Region**

Pendra tehsil is the only tehsil comes under this group predominance of Bhata soil in responsible for lower productivity.

**Moderate Carrying Capacity Area**

Two tehsils were found in this category. They include Kota and Bilha. Between these two tehsils, Bilha has recorded highest (7.5 persons/ hectare/year) carrying capacity (Fig. 2). Kota tehsil is covered with forested area and rugged terrain and poor Bhata soil providing moderate carrying capacity.

**High Carrying Capacity Area**

Masturi and Takhatpur tehsils showed high carrying capacity with 8.7 and 8.3 persons/hectare/annum. High carrying capacity of the area is because of availability of irrigation facility from Khanang Medium Irrigation Project, supplemented by plain fertile land.

**Very High Carrying Capacity Area**

Lormi and Mungeli tehsils enjoy the highest carrying capacity (more than 9 persons/hectare/year) in the study region. Maniyari and Agar rivers, Canal irrigation, black soil provide favourable condition for high production.

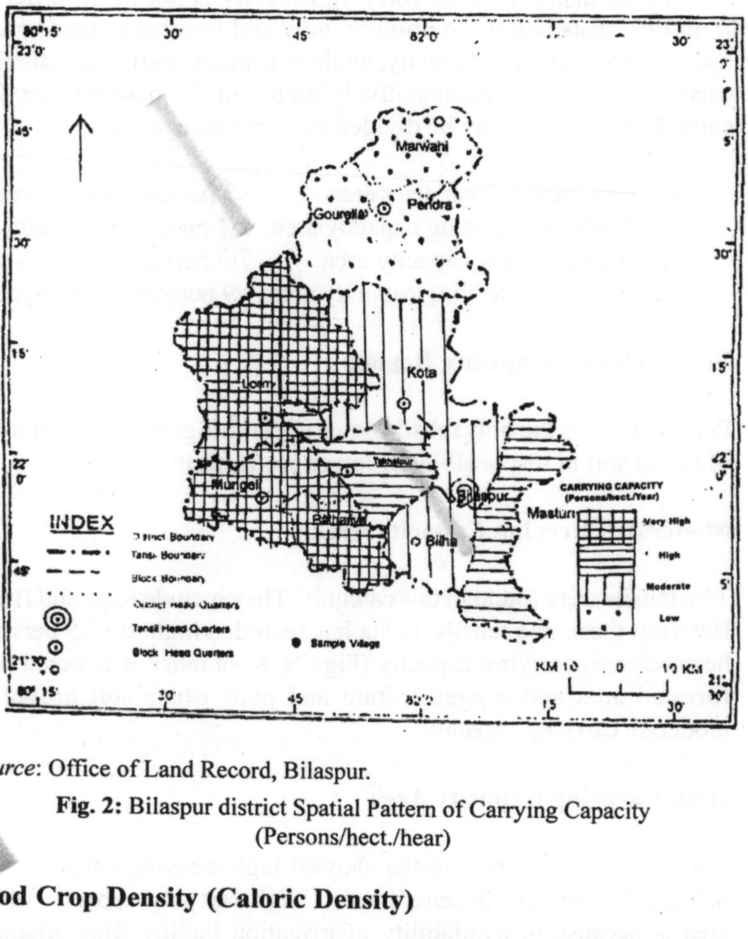


*Source*: Office of Land Record, Bilaspur.

**Fig. 2:** Bilaspur district Spatial Pattern of Carrying Capacity (Persons/hect./hear)

## Food Crop Density (Caloric Density)

For the comparative study of region, variation in carrying capacity food crop density, is a good measure. In other words, food-crop density is defined as the density of population per unit area under food-crop. In the present paper, population in tehsil is divided by area under paddy crop.

In the study area food-crop density is 6.3 person/hectare/year. It ranges 12.4 (maximum) person/hectare/year in Bilha to 4.20 (minimum) person/hectare/year in Masturi while in Kota, Pendra it is 5.5 and 6.2 respectively (see Fig. 3). In Mungeli Tehsil it is comparatively high, i.e. 9.0 person/hectare/year.

**Table 1:** Population Food available in Bilaspur District.

| *Sl. No.* | *Name of Tehsil* | *PPU C/H/Y in million* | *PPU/ SNU* | *Food Crop Density* | *Surplus/ Deficit* |
|---|---|---|---|---|---|
| 1. | Bilha | 79.5 | 7.5 | 12.4 | – 4.9 |
| 2. | Bilaspur | 76.2 | 7.4 | 12.4 | – 5 |
| 3. | Masturi | 91.5 | 8.7 | 4.20 | + 4.5 |
| 4. | Takhatpur | 87.6 | 8.3 | 5.16 | + 3.14 |
| 5. | Pendra | 59.1 | 5.0 | 5.5 | – .5 |
| 6. | Kota | 78.7 | 7.4 | 6.2 | + 1.2 |
| 7. | Mungeli | 97.2 | 9.2 | 9.0 | + .2 |
| 8. | Lormi | 99.2 | 9.4 | 6.1 | + 3.3 |
| 9. | Bilaspur District | 81.0 | 7.7 | 6.3 | + 1.4 |

*Source*: Land Record Office, Bilaspur.

## Population: Food Balance

Population pressure is the appropriate index of agriculture development on an area. For this comparison of carrying capacity and food-crop density has been made. This study helps to identify the area with surplus carrying capacity and deficit carrying capacity.

The study has shown that the study area as a whole has an average carrying capacity is found. These are Bilha, Bilaspur and Pendra Tehsils. In Bilha and Bilaspur Tehsils, due to increasing urbanization available agriculture land decreased. Bilaspur is a urban centre in the district so the density of population is comparatively higher than any other area.

In Pendra tehsil food-crop density is 5.5 person/hectare/year and carrying capacity of land is 5.0 persons/hectare/year leading to deficit. Undulating terrain, lack of irrigation facility and other infrastructure facilities are responsible for such a scenario.

In Mungeli and Kota tehsils, very small surplus carrying (+ 1.2 persons/hectare/year in Kota, + 0.2 in Mugeli) was noticed. Remaining 3 tehsils enjoy surplus carrying capacity ranging 3.14 persons/hectare/ year at Lormi to 4.15 persons/hectare/year at Masturi.

## Conclusion

Overall, in the study area, 62.5 per cent of tehsils enjoyed surplus carrying capacity, but the sufficiency is under marginal. In the district, food-crops must be increased by taking necessary steps for improving

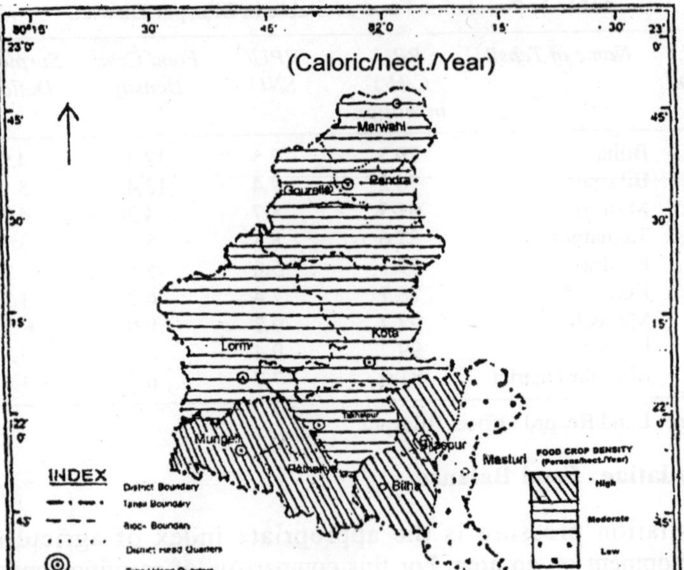


Fig. 3: Bilaspur District: Food Crop Density

*Source*: Office of Land Record, Bilaspur.

the carrying capacity of land especially in northern part of the district. There is enough scope for increasing the productivity of agriculture in the region by using HYV, fertilizers, and modern equipments providing agricultural extension facilities, with farm-based education, the carrying capacity can be further improved.

## REFERENCES

Gatade, D.G. (2004): "Spatial Pattern of Carrying Capacity of Land in Satara District/Maharastra State", *Deccan Geographer*, Vol. 2. No. 2, pp. 11-18.

Kashyap, D.D. (1993): *Agricultural Productivity and Carrying Capacity of Land on Raigarh District*", Unpublished Ph.D. Thesis, G.G.D. University, Bilaspur (C.G.).

Reddy, M.V. & Reddy, N.B.K. (1993): "Spatial Pattern of Carrying Capacity of Land in Cuddapah District", *Transactions*, Vol. 15, No. 1, pp. 13-14.

# 19

# Land Use Changes in the Urban Vicinity
## *A Micro-Level Study*

Sushma Bomjan and Mrityunjay M. Jha

### Introduction

The optimal utilisation and management of natural resources for sustainability to maintain equity between the demands of the present and the needs of the future generation was the main objective of the Rio-conference of 1992. The World Summit on Sustainable Development held at Johannesburg in 2002, identified five major areas of concern, viz. Water, energy, health, agriculture, biodiversity and ecosystem management. Out of these, water, agriculture and biodiversity, clearly highlight the importance of land use and land cover. This generated a never ending debate on 'sustainability'. The exploitation of natural resources is dynamic and it should be seen *vis-à-vis* population. Land, being the most important natural resource is easily influenced by population growth and technological advancement.

The pressure generated by population growth reflects two key demographic factors: the rate of population growth and the absolute number of people added each year. The rapid growth in human numbers, plus their distributional patterns induces critical constraints on the development process as well as unsustainable burden on the environmental resource base. In developing countries, importance of population growth *vis-à-vis* development has been recognised by planners and policy makers. Therefore, for a strong and sustained economy to pursue development programmes, up-to-date land resources data becomes inevitable.

Remote sensing because of its synoptic view and repetitive coverage, provides useful information on land use dynamics. The changes in land use due to natural and anthropogenic activities can be observed using current and old remote sensing data. The fast breeding 'geographical information system' is a powerful set up tools for collecting, retrieving and transforming large data base. It is system for handling spatial and non-spatial data on natural resources and socio-economic attributes. The integration of remote sensing and GIS has been used in several comprehensive studies on land use. The GPS/GIS provide means to collect and integrate the information on land resources derived from remote sensing data with other collateral information to draw appropriate action plans for sustainable development of land water resources (Gautam *et al.*, 2003).

Gautam, Raghavaswamy, Nagaraja and Ravi Sankar (1994) dealt with the utility of satellite remote sensing for survey, mapping and monitoring of land use. Nagaraja and Gautam (1994) in another attempt focused on GIS for land use mapping and environmental management. On land resources theme, some noteworthy contributions are on land suitability and mapping (Chaurasia, 1996; Kam, 1995), change detection (Somparan, 1995; Pierre, 1992; Trung 1993), and integrated remote sensing and GIS technologies in Rural land use and Development (Wilkinson, 1996; Kushwaha *et al.*, 1996).

**Objectives of the Study**

The main objectives of this paper are:

1. To identify the overall land use changes and highlight the significance of the agricultural land use changes.
2. To probe the role of population growth in land use changes, specially on agriculture and forest.

**Scope of the Study**

The main scope of this study lies in the analyses of remotely sensed data, identifying land use changes in GIS environment. Integration of visual interpretation and non-spatial data using GIS technique is aimed at preparing a database, which in turn would suggest remedial guidelines. This study area was selected with spatial consideration to the availability of satellite imageries. The study is limited within the part of Western Doon valley near the urban fringe of Dehradun city.

**Physical Characteristics of the Study Area**

The study area is a part of Western Doon valley located in the proximity of Dehradun city. It covers an area of about 120 sq. km. extending from 30′17″N to 30′22″N and 77′50″E to 77′57″E. Alluvial plain is main physiographic unit comprising river terraces and flood plains—Asan, Tons, Song and their tributaries, namely Bainkhala, Suvarna and Barawala. The maximum temperature goes up to 35°C and minimum 1°C. Average rainfall ranges between 1750-2000 mm. Mainly three seasons are witnessed in Doon valley—summer (April—mid-June), rainy (mid-June—mid-September) and winter (October—March). Rainfall may not follow the rigid classification. Hilly and piedmont soil constitutes the maximum area. In the middle part old alluvium and recent alluvium soils with sandy loam texture are also found. The socio-economic condition of the villages is fairly good. Agriculture is the main stay of the economy. Dominant villages of this study area are Accadia Grant, East Hope Town, Hassanpur, Sherpur, Kandhauli, Keshouala, Shudhonwala and Shishambara.

**Data Sources**

Data were collected and developed in a manner that allows comparison. Data acquisition was made from various sources:

- *Remote sensing data*: SPOT-1 (XS) FCC data of November 1987 and IRS-1D LISS-III FCC of October 1998 were taken to gauge change in the land use pattern.
- *Census data*: Village and Town Directory of 1981 and 1991 along with District Census Hand Book of Dehradun district, was used to get data on Population, Agriculture and Forest. All these were obtained from the National Informatics Centre (NIC), Dehradun.
- *Other Ancillary data*: It was essential and basic to use topographic sheet (53 F/15) of the area at 1: 50,000 scale. Physiographic soil map of Doon valley was taken from Agriculture and Soil Division of IIRS, Dehradun. Other information was collected from the field visits on nature of past land use with observed changes.

## Methodology

The methodology for the present study is divided into three parts:

### *Preliminary Interpretation*

A base map was prepared, extracting its boundary from the Survey of India (SOI) topo sheet at 1: 50,000 scale showing important features. The boundaries of the villages within the extent of the study area was traced to prepare different themes.

Visual interpretation technique was used to study the digitally enhanced products on the basis of image characteristics. This technique was chosen to avoid problems of mis-classification in digital image analyses which are known to be caused by spectral confusion. The image features on the imageries were interpreted for land use classification using the appropriate interpretation keys, to enable direct comparison and detect change.

Field visits were conducted for ground checks. The corrected maps were scanned, rasterised and converted into digital maps. These maps were later geo-referenced by resampling using ERDAS/IMAGINE image processing software. The land use classification system was based on interpretation keys as well as indicators from the field. The main objective of field study was to check the ground truth with the base map, which was prepared by visual interpretation. And also to find out the regions responsible for receding forest areas, population expansion and changing agricultural pattern.

### *Spatial and Non-spatial Data Generation*

All thematic layers were prepared by on-screen digitization using ARC/VIEW and further taken to ARC/INFO to generate various vector maps. Non-spatial data on population, agriculture and forests were created as a database and added to the village maps as attributes.

### *GIS Analyses*

GIS technology has emerged in the last two decades, providing rapid analyses, consisting of overlaying, crossing and buffering techniques.

To find out landuse change, change detection technique based on two date classification was used. This method has been effectively used in several studies. Overlaying is a wonderful method in finding changes for establishing relationship between the two data sets. Change detection involves the use of multispectral data sets to discriminate areas of land cover change between dates of imaging. This technique is very useful in diverse applications, as land use change analysis, monitoring of urban expansion, shifting cultivation, shifting of river course, deforestation, post-disaster damage assessment and other changes. Buffering enables to know the latent influence of a place or an area.

## Overall Land Use Changes

Overlaying allows to gauge the change, and it is the process of identifying the differences in the state of an object or phenomenon. Remotely sensed data is widely used for instant interpretation as well as for change detection using repetitive coverage at short intervals. Land use maps of 1987 and 1998 (Figures 1 and 2) generated from the images were classified into 9 categories, and further these two were crossed to gauge change in inter-class land use change plus overall change. Through this, the dynamics of land use change can be witnessed. Changes in different land use classes are shown in Table 1. There is a clear-cut decline in the forest area in all its sub-classes. This shows a declining trend and is a matter of concern. All three forest sub-classes together constituted 57.5 per cent of the total land in 1987 whereas, in 1998 it came down to 45 per cent. Cross tabulation analysis shows that most of the losses went to agriculture. Field visits suggest that vegetation disturbances vary from partly removal of trees to clearings and complete tree felling for agricultural purposes.

## Agricultural Land Use

There is steady increase in agricultural land over the years. Highly intensive agriculture land has not changed much. The maximum change is visible in the moderately intensive agriculture. Though, low intensive agriculture still occupies the maximum agricultural area. In the moderately cultivated land, there has been addition of around 10 per cent of land from other classes, mainly low intensive agriculture

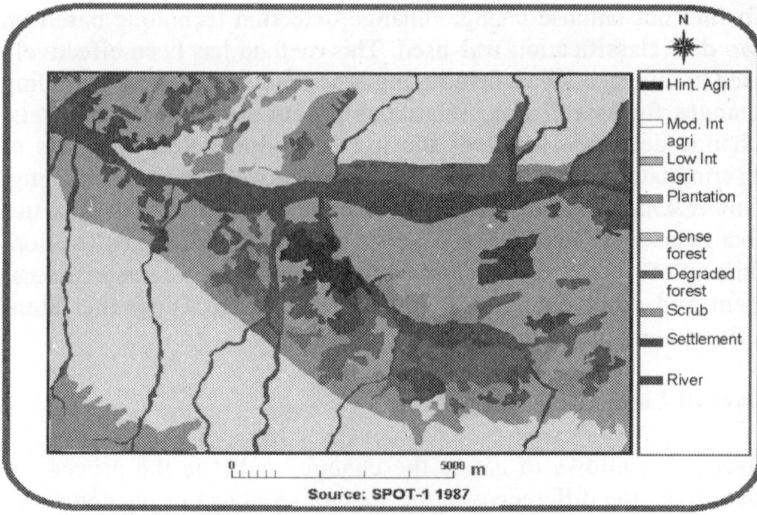


**Fig. 1:** Land use of Part of Western Doon Valley 1987.

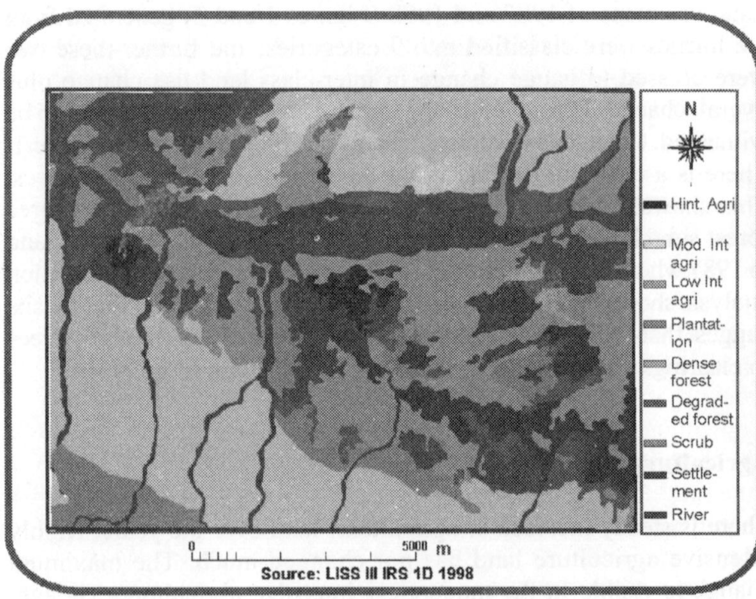


**Fig. 2:** Land use of Part of Western Doon Valley 1998.

**Table 1:** Land Use over the Years in Part of Western Doon Valley (area in %).

| *Land use Class* | *1987* | *1998* |
|---|---|---|
| Highly Intensive Agriculture | 1.24 | 1.27 |
| Medium Intensive Agriculture | 4.08 | 12.41 |
| Low Intensive Agriculture | 18.4 | 16.96 |
| Plantation | 4.82 | 4.67 |
| Dense Forest | 36.77 | 33.3 |
| Degraded Forest | 11.42 | 7.4 |
| Scrub | 9.64 | 4.48 |
| Settlement | 4.24 | 8.97 |
| River | 9.39 | 10.54 |
| **Total** | **100** | **100** |

and scrubs (4.49 and 2.95 per cent respectively). And from this category 3 per cent has been taken by others mainly low intensive agriculture and settlement.

Striking changes in different land use classes were found over the years. This was done by crossing the village boundary map and overall land use change as per satellite data. In Sherpur village, highly intensive agriculture has started which did not exist earlier and in East Hope Town village experienced tremendous increase in moderately intensive agriculture. Hasanpur showed remarkable growth of more than 100 per cent in population, which is due to immigration and expansion of agricultural land. Most of the villages have experienced growth in population and agriculture. In these villages, forest area has declined.

The growth of civilizations and population growth have led to the expansion of agricultural land and settlement area. Settlement areas which constituted only 4.45 per cent of the total land in 1987 went up to 9.28 per cent in 1998. Gain was mainly from low intensive agricultural land and scrubs. This indicates the expansion or movement of settlements towards forest areas. Location of all these villages in the southern part of the study area which is closer to the forest area is significant. Increase in the settlement area in the western part could be noticed easily. Expansion in the settlement area in Hasanpur village has been cross checked with the increase of population.

### Conclusions and Recommendations

Land use constitutes an essential element in planning and management of land resources. The result of the above case study indicates that there has been change in land use/land cover. Some changes are negative like declining dense forest cover area. Some changes seemed to be inevitable, as with the growth of population the agricultural land needed expansion.

Applicability of the integrated GIS and Remote Sensing data has acquired much significance which is getting reflected from this study. It has shown significant results and can be put on record. Certain inevitable conclusions have been drawn here. Integration of spatial and non-spatial data has been very useful, hence, the much needed multi-disciplinary approach may be important in future studies. Increasing human population is responsible for change in land use. Population and agricultural expansion near the forest area is significant.

The study area is experiencing the chain of degradation from dense forest to degraded forest to scrub and to low intensive cultivation. Here, fuel wood requirement and need for agricultural land are reasons for depleting forest area. Therefore, reserve forest located in the North and South of the study area need to be monitored regularly.

## REFERENCES

Census of India. (1981): *Village and Town Directory*. Dehradun

Census of India. (1991): *Village and Town Directory*. Dehradun.

Census of India. (1981): *District Census Handbook*, Dehradun.

Carvalho, Elra. E. (1997): The Transformation of Land Use (1984/94) in Western Part of Rio De Janeiro, Brazil—An Application of Remote Sensing, in (ed.). I.V. Muralikrishna, *GIS and RS Applications*. Allied Publishers. New Delhi.

Chaurasia, R. *et al.* (1996): Land Use Change Analyses for Agricultural Management—A Case Study of Tehsil Talwandi Sabo, Punjab. *Journal of Indian Society of Remote Sensing*. Vol. 24, No. 2, pp. 115-123.

Gautam, N.C. *et al.* (1994): Satellite Remote Sensing for Survey, Mapping and Monitoring of Land use/Land cover: A few examples. (ed.) N.C. Gautam, V. Raghavaswamy and R. Nagaraja. *Space Technology and Geography* NRSA, Government of India.

Gautam, N.C. *et. al.* (2003): *Land use/Land cover and Management Practices in India*. BS Publications, Hyderabad.

Kam Tin-Seong. (1995): Integrating GIS and Remote Sensing Techniques for Urban Land Cover and Land Use Analysis. *Geocarto Journal*. Vol. 10, No. 1, pp. 39-49.

Kushwaha, S.P.S. *et al.* (1985): Interfacing Remote Sensing and GIS Methods for Sustainable Rural Development. *International Journal of Remote Sensing*. Vol.17, No. 15, pp. 3055-3069.

Nagaraja, R. and Gautam, N.C. (1994): GIS for Land Use Planning and Environmental Management. *Space Technology and Geography* (ed.). N.C. Gautam *et al.* NRSA, Government of India.

Pierre, Vincent. (1992): The Development of a Land Use Mapping Methodology for Northern Thailand. *Asia-Pacific Remote Sensing Journal*, Vol. 5, No. 1, pp. 61-68.

Rao, U.R. (1991): Remote Sensing for Sustainable Development. *Journal of Indian Society of Remote Sensing*. Vol. 19, No. 4, pp. 117-135.

Somparn, Sangawongse. (1995): Landscape Change Detection in a Tropical Monsoon of Northern Thailand Using LANDSAT™ Data: Applications in Chpang Mai Basin. *Asia-Pacific Remote Sensing Journal*, Vol. 8, No. 1, pp. 59-69.

Trung, Lyong Phang (1993): The Detection of Land Use/Land Cover Changes Using Remote Sensing and GIS in Vietnam. *Asia-Pacific Journal of Remote Sensing*. Vol. 15, No. 2, pp. 62-65.

Wilkinson, G.G. (1996): A Review of Current Issues in the Integration of GIS and Remote Sensing Data. *International Journal of Remote Sensing*. Vol. 10, No. 1, pp. 85-101.

# Epilogue

The papers in this book bring out significant dimensions related to monitoring land use changes, planning and proper management of land resources. Management of land is so fundamental to sustainability that it should permeate the whole fabric of development. This can be achieved through a very close interaction between scientists, planners and civil society. Some of the papers also throw up the need for a national land use policy based on improved information on land use. Until information on land use is improved, development planning will remain without an adequate basis on major issuers, including natural resource management. For achieving sustainable development at the grassroots level, knowledge and views must be shared between planners, advisers and the people who manage land resources, because ultimately it is people who manage land resources, i.e. farmers, small entrepreneurs and other land users.

There is a dire need of proper human resource development by providing employment opportunities and suitable land use planning. The ability to forecast land use change and to forecast the consequences of change will depend on our ability to understand the past, current and future drivers of change. Besides, emerging social and political factors may also have significant impact on the future land use. Hence, we can say that the patterns of land use changes and land management are the result of economic, environmental, political, social and technological interactions. Such interactions would necessarily require more empirical studies that would combine micro-level case studies and regional/global level studies, involving remote sensing at different scales, statistical analysis and participatory appraisal. Under any given situation and circumstance, biophysical forces and anthropological pressures will remain in the core of all research studies. At the crucial juncture of human civilization, we require multidisciplinary cooperation for arriving at significant intersections, viz. social, economic, ecological and technological impacts on land use changes *vis-à-vis* agriculture, water resources, land management and forestry.

We need Policy Guidance, which should be a mix of recommendations from the experts, planners and policy makers and that of the bottom up messages emanating from the grassroots level, spelling out best practice in developing and operating strategic processes for sustainable land use, and how best the developing countries can get leverage in such processes, and include a set of principles which underpin the development of effective strategies in many developing countries.

# Index